K. BUGAYEV, Y. BYCHKOV,
Y. KONOVALOV, V. KOVALENKO,
E. TRETYAKOV

IRON AND STEEL PRODUCTION

Translated from the Russian
by
Ivan V. Savin

Books for Business
New York-Hong Kong

Iron and Steel Production

by
K. Bugayev
Y. Konovalov
et al.

ISBN: 0-89499-109-4

Reprinted from the 1971 edition

Books for Business
New York - Hong Kong
http://www.BusinessBooksInternational.com

The Russian Alphabet and Transliteration

А	а	a	К	к	k	Х	х	kh
Б	б	b	Л	л	l	Ц	ц	ts
В	в	v	М	м	m	Ч	ч	ch
Г	г	g	Н	н	n	Ш	ш	sh
Д	д	d	О	о	o	Щ	щ	shch
Е	е	e	П	п	p	ъ		"
Ё	ё	ë	Р	р	r	ы		y
Ж	ж	zh	С	с	s	ь		'
З	з	z	Т	т	t	Э	э	e
И	и	i	У	у	u	Ю	ю	yu
Й	й	y	Ф	ф	f	Я	я	ya

Chemical Symbols

Al	aluminium	H	hydrogen	Pb	lead
Au	gold	Mn	manganese	S	sulfur
B	boron	Mo	molybdenum	Si	silicon
C	carbon	N	nitrogen	Sn	tin
Co	cobalt	Nb	niobium	Ti	titanium
Cr	chromium	Ni	nickel	V	vanadium
Cu	copper	O	oxygen	W	tungsten
Fe	iron	P	phosphorus	Zn	zinc
				Zr	zirconium

Contents

Contents

Introduction

Metals and alloys are the chief materials for the manufacture of various mechanisms, machines, and constructions in modern engineering. This is due to the fact that metals and metallic alloys in particular, are capable of providing any properties desired in a piece or structure, such as strength, machinability, wear resistance, corrosion resistance, and others.

The metals widely used are iron, aluminium, nickel, lead, zinc, tin, and others. In recent years, titanium, beryllium, zirconium, and a number of other metals have been gaining a wide recognition.

Pure metals lack in many properties essential in engineering applications. This is why alloys, obtained by melting together two or more metals, are used on a large scale. These alloys may contain non-metallic constituents, such as carbon.

Characteristic of alloys, along with the general metallic properties (e.g., lustre), are their special properties. There are, for example, alloys with a high resistance to acids or, say, with a high magnetic permeability.

Most widely used are iron and iron alloys, their applications ranging from the minutest radio engineering components to modern ocean-going ships, from small household ware to the largest engineering projects.

Iron is almost never used in pure state. Chief industrial materials are iron alloys, the so-called *ferrous metals*. Iron alloys usually contain carbon, silicon, manganese, sulfur, and phosphorus. Special properties may be imparted to alloys by the addition, in the process of manufacture, of alloying elements, such as chromium, nickel, molybdenum, aluminium, titanium, tungsten, vanadium, copper, and others.

The properties of iron alloys are greatly affected by their carbon content. In terms of carbon content, iron alloys are subdivided into cast iron, steel, and pure iron.

Cast iron is an alloy with more than 2% C.

Steel is an alloy of iron and other elements, containing less than 2% C.

Very mild steel which contains less than 0.04% C is called *commercial iron* or, for short, iron.

Common grades of cast iron contain 3.5-4.5% C, while steel (upwards of 400 grades of steel are being currently manufactured) generally contains not more than 1.4% C.

Steel possesses many properties of a greatest value. It has high strength and toughness; when heated, it can readily be shaped, i.e., forged, stamped, or rolled. Special properties, such as acid resistance, heat resistance, etc., may be imparted to steel.

The widest industrial use of steel and cast iron is also due to the fact that iron is one of the most frequently occurring elements in the earth's crust and that it is readily extracted. The fraction of iron in the mass of the upper earth strata is 4.6 per cent, and in this respect it is inferior only to oxygen, silicon, and aluminium.

Vast reserves of iron ores, their relatively easy mining, and the simple process of iron extraction from the ores (as compared to the manufacture of other metals) put iron and steel to the forefront among the materials employed in national economies.

Manufacture of iron and steel and their working involve the following main stages:

1. Mining and preparation of iron ore.
2. Smelting of iron from iron ore with the use of auxiliary materials performed in blast furnaces.
3. Processing of smelted iron to steel in converters, open-hearth or electric furnaces.
4. Hot working of steel ingots to required shape, dimensions, and properties by various rolling, forging, stamping, pressing, drawing, and heat-treatment techniques.

Widely practiced in mechanical engineering is casting of finished parts from cast iron or steel.

REVIEW QUESTIONS

1. What materials are called *ferrous metals?*
2. Describe the overall sequence of iron and steel manufacture.

SECTION I • FUNDAMENTALS OF PHYSICAL METALLURGY

Physical metallurgy is the science studying the composition, structure, and properties of metals and alloys and the interplay of these factors. Knowledge of the relationship between the composition and structure of a metal, on the one hand, and its properties, on the other, allows creation of metallic alloys with predetermined properties and a fuller use of the properties of metals in structures.

The properties of metals may be conventionally divided into physical, mechanical, and working properties. Many properties of metal are interrelated, but it aids in assessing the behaviour of a metal during both manufacture and service of a product.

The physical and mechanical properties of metals are atomic mass, density, heat conductivity, electric resistivity, melting point, heat capacity, coefficient of linear expansion, magnetic properties, hardness, strength, plasticity, toughness, brittleness, elasticity, endurance, creep, etc.

The physical and mechanical properties of metals and alloys depend not only on the nature of atoms composing the material, but also on the strength of the bonds between the atoms. These properties should be taken into account when choosing a metal or an alloy for specific applications. Physical and mechanical properties of pure metals are given in reference books. It should be noted that the properties of metallic alloys may differ greatly from the properties of the constituent metals.

However, the knowledge of the physical and mechanical properties only is not enough to assess the behaviour of a metal during the manufacture and the service of a piece.

Therefore, the suitability of a metal for various applications should be evaluated with due regard for its *working properties*, such as forgeability, weldability, flowability, stamping properties, machinability, corrosion and wear resistances, and other properties.

The working, physical and chemical properties are closely interconnected. For example, stamping properties depend on the hardness, plasticity, and other properties of the metal involved. The working properties, determined by means of special tests, give a good idea of a metal from the standpoint of its behaviour when subjected to rolling, machining, and other operations in the production processes.

CHAPTER 1

Mechanical Testing of Metals

Mechanical tests are necessary to estimate the ability of metal to resist mechanical failure or to endure stresses over prolonged periods of time.

Mechanical tests should be conducted in strict compliance with the respective standards. For example, in the USSR tests for tensile strength should comply with the Soviet State Standard GOST 1497-61, and tests for Brinell hardness, with GOST 9012-59. These standards prescribe the shape and dimensions of test specimens, methods for measuring or calculating the characteristic to be found, the load to be applied, and other relevant information. Standardization of testing procedures provides a firm basis for comparison of results obtained in different testing laboratories.

Depending on the rate of load application, the mechanical tests fall into three groups: static, impact, and endurance tests.

In static tests, the load acting on the specimen is caused to change slowly and gradually, and, accordingly, the rate of specimen deformation is small. This allows a sufficiently accurate measurement of the specimen deformation throughout the test.

The results of the test may be reported in terms of the following characteristics:

1. Magnitudes of *forces* (N) or *stresses* (N/m²), which may allow a judgement on the resistance of the specimen metal to deformation in the course of the test or at the time of its failure.

2. Absolute (mm) or relative (%) values of *elastic and residual deformations* caused by known forces.

3. Values of full (N·m) or specific (N·m/m²) *work of deformation* against the resistance of the specimen metal.

Static tests (in tension, bending, etc.) are widely used at production plants, but they provide no basis for estimating the capability of the metal to resist impact or fluctuating loads.

In impact tests*, the specimen is loaded instantaneously, and the rate of deformation is very high.

The following characteristics may quantify the results of impact tests:

1. Full or specific *work of dynamic deformation* of the specimen metal.

2. Absolute (mm) or relative (%) *deformation* of the specimen in the test.

3. Mean or maximum *load* on, or *stress* in, the specimen during the test.

Impact tests give an indication of the metal brittleness under impact loads, including that at different temperatures. The metal is embrittled by notching the specimen so that stresses are concentrated at the notch. In real pieces or structures, a similar stress concentration appears at sudden transitions from one section to another, cracks, rough machining marks, large foreign inclusions. All these flaws increase the brittleness of metal and may cause rapid failure.

In endurance tests, checked is the capacity of the metal to resist cracking due to metal fatigue under repeated alternating loads. The loads in these tests are usually small, however, the number of test cycles (i.e., the number of times the load is applied) may be as high as a few millions.

---

* Also known as Charpy or V-notch tests.

Tensile tests. Mechanical tests are performed on special specimens (Fig. 1). The dimensions and the shape of specimens should comply with the relevant specifications, so that properties of a grade of steel after different heat treat-

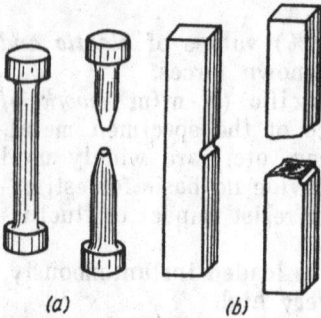

Fig. 1. Specimens for mechanical testing (left, before testing; right, after failure)

(a) tension test specimen; (b) impact test specimen

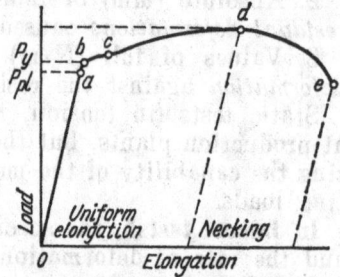

Fig. 2. Tensile stress-strain diagram

P_{pl} — load at proportional limit;
P_y — load at yield

ments or of different grades of steel after a similar heat treatment may be compared.

Most generally, a metal is subjected to a static tensile test in special machines. The test specimen is either flat or cylindrical (Fig. 1a); in case of a wire, the test specimen is a specified length of wire.

The specimen in the machine is stretched to rupture. The force on the specimen grows gradually and at a certain load the specimen begins to elongate and change shape. There is a definite relationship between the load and the elongation. The values of load and elongation are recorded by a special built-in device to give a tensile stress-strain diagram, i.e., a load-*vs.*-elongation curve.

Fig. 2 illustrates a typical tensile stress-strain diagram for mild steel.

At the beginning of loading, the elongation grows in direct proportion to the load (straight line O*a*). The maximum stress at which the strain-stress dependence still remains linear is the *proportional limit* calculated as:

$$\text{Proportional limit} = \frac{\text{Load at proportional limit, N}}{\text{Original cross-sectional area of specimen, m}^2}$$

If the stress in the specimen is less than proportional limit, the deformation is elastic and disappears as soon as the load is removed.

If the stress exceeds proportional limit, the linear relationship no longer exists between the load and the deformation, and the latter remains after the load is removed. On the tensile stress-strain diagram, the straight line 0a passes into curve ab, and a plateau may occur on it at a certain load. This means that the deformation of the metallic specimen grows with no increase in the load. The stress at which the metal deforms without appreciable increase in the load is designated as its *yield point* and determined as:

$$\text{Yield point} = \frac{\text{Load at yield, N}}{\text{Original cross-sectional area of specimen, m}^2}$$

If there occurs no plateau on the tensile stress-strain diagram, the yield point is the stress that causes a residual elongation of 0.2 per cent.

A further increase in the load causes elongation (curve cd) and necking (curve de) and brings about the failure of the specimen.

The maximum stress which the metal of the specimen can resist without failure is the *ultimate (tensile) strength*:

$$\text{Tensile strength} = \frac{\text{Maximum load, N}}{\text{Original cross-sectional area of specimen, m}^2}$$

Elongation of the specimen in the tensile tests may serve as a measure of the metal plasticity; the percentage elongation is determined as:

$$\text{Percentage elongation} = \frac{\text{(Final length)} - \text{(Original length)}}{\text{Original length}} \times 100\%$$

Percentage reduction of the original cross section is determined in the same manner as the percentage elongation.

Impact tests. In impact tests, a notched specimen (Fig. 1b) is supported in an impact testing machine, and a pendulum strikes on the specimen side opposite to that with the notch. The pendulum fractures the specimen and rises inertially to a certain height. The difference in heights before and after the impact is a measure of the energy absorbed in fracturing the specimen. The value of notch toughness is found as:

$$\text{Notch toughness} = \frac{\text{Energy absorbed in fracturing, N} \cdot \text{m}}{\text{Cross-sectional area under notch, m}^2}$$

A brittle metal fails easily under impact, very little deformation of the specimen occurring at the point of failure; a plastic, tougher metal fails at higher impact loads, and the specimen is very much deformed at the point of failure.

Hardness is the capacity of a metal to resist penetration by a harder body.

Hardness is determined by pressing a hard indenter into the metal under test. There are also other methods for measuring hardness (Vickers, Shore, etc.), the use of which is dependent upon the dimensions of the specimen or piece, the expected hardness of the metal, etc.

Brinell test. The test is performed in a special tester by pressing a hardened steel ball of a specified diameter (2.5, 5, or 10 mm) into the specimen by a specified load (15-3,000 kgf) during a predetermined period of time (10-60 s). The diameter of the impression is then measured and the hardness number calculated as:

$$\text{HB} = \frac{\text{Load, kgf}}{\text{Impression area, mm}^2}$$

Practically, the hardness of a metal is determined by measuring the impression diameter and finding the corresponding hardness number from a table.

The harder the metal under test, the greater should be the load. A harder metal is usually the stronger one.

The Brinell method is not applicable to very hard steels (with HB higher than 450 kgf/mm^2), neither to thin pieces or specimens.

Rockwell test. The test consists in indentation of the specimen with a diamond cone or a steel ball by applying two loads in succession: minor (0.1 kN) and major (0.6, 1, or 1.5 kN).

The depth of the indenter penetration into the specimen under the action of these loads characterizes the hardness of the metal under test. The harder the metal, the smaller is penetration of the diamond cone (or a steel ball which is used for softer metals).

The Rockwell hardness number is indicated in conventional units by a pointer of the hardness tester.

The Rockwell method is applicable to harder and thinner specimens or pieces; the impression obtained is small in size.

Metal fatigue. In static tensile tests, the specimen is brought to failure by applying stresses exceeding its ultimate strength. However, if the load (and, therefore, the stress) is a varying one and is repeated a great number of times, the piece or specimen may fail under considerably lower stresses, no deformation of the specimen or piece being observable.

This failure is caused by metal fatigue. The *endurance* of the metal, i.e., its capacity to withstand a great number of repeated fluctuating stresses, is a very important property since most of machine elements operate under alternating loads.

Endurance tests are carried out on special machines in which a specimen is subjected to loads varying in magnitude and direction, such as compression and tension; torsion and bending; repeated impacts, etc. In these tests, a relationship is established between the load value and the number of cycles the specimen stands before failure; the stress which the specimen withstands during a predetermined number of cycles is referred to as the *endurance limit*. The number of cycles can be very significant; for example, for steel it is 10 million.

Determination of mechanical properties of metals is essential in many cases. A designer, when devising a piece, should choose a metal of suitable properties to enable the piece to operate for a long time without failure.

Even a small residual change in dimensions and shape is not allowed in machine parts. Therefore, the designer takes care to select a metal of such a yield point or proportional limit which will keep the strains within the elastic range, this being achieved by calculating the stresses arising in the piece during its operation. Naturally, the yield point of the material involved should be chosen well above the operating stresses, otherwise the piece will immediately be distorted and fail.

Mechanical properties of steel are tested at the iron and steel plant so as to allow engineering works, the main consumers, to use the steel suitably.

Heat treatment, such as quenching, greatly affects the mechanical properties of steel, and because of this the properties are tested after every cycle of heat treatment to which a piece is subjected in the process of its manufacture. Frequently, a hardness test is sufficient.

Mechanical properties of a great many steels are specified by standards or specifications which are compulsory for manufacturers of metal products.

CHAPTER 2

Structure of Metals and Alloys

2.1. Crystals of Metals

In contrast to amorphous solid bodies whose atoms are distributed irregularly in space (one example is glass), metals are crystalline bodies, and their atoms are arranged

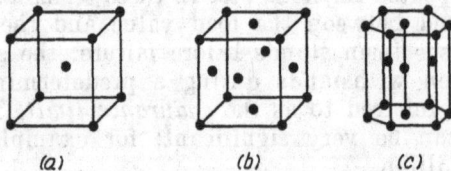

(a)   *(b)*   *(c)*

Fig. 3. Crystal lattices of metals

(a) body-centred cubic (iron at temperatures below 910 °C and above 1,400 °C, chromium, molybdenum); (b) face-centred cubic (iron in the 910-1,400 °C range, lead, copper); (c) hexagonal (zinc, magnesium)

in geometrically orderly manner at predetermined distances from one another.

The location of atoms in different metals may be illustrated with the aid of a unit cell.

Figure 3 shows common types of crystal lattices of metals. It is obvious that in the lattices of different types the number of atoms per unit volume, the interatomic distances and the number of neighbours of any given atom are different, too. This affects both the properties of pure elements and their mutual solubility in alloys.

Atoms are located at lattice points; the distance between two adjacent atoms is called the *lattice constant*. Atoms of different metals differ in size, but still the metals can dissolve each other. Therefore, the lattice structure of metal *A* changes if a given amount of metal *B* is dissolved in it.

If metal *A* dissolves some quantity of metal *B*, the atoms of the latter may be located either at lattice points instead of the atoms of metal *A* (substitutional solid solution) or between the atoms of metal *A* (interstitial solid solution). In both cases, the interatomic distance in the solid solution differs from that in the solvent, i.e., the lattice constant changes. Generally, the atoms in metals are spaced at 0.000,000,2 to 0.000,000,6 millimetre.

A phenomenon inherent in many metals is polymorphism or allotropy, which consists in that the type of the crystal lattice and its constant change with the surrounding conditions. For example, iron at temperatures up to 910°C has a body-centred cubic lattice (alpha-iron), and in the 910-1,400°C range it has a face-centred cubic lattice (gamma-iron). At a higher temperature the lattice again becomes body-centred cubic (delta-iron), but with a larger interatomic distance than in alpha-iron.

The re-arrangement of the lattice is accompanied by a change in the properties of iron, such as strength, plasticity, capacity for dissolving other elements, etc.

A real metal is polycrystalline, i.e., it consists of a great number of small crystals, or grains. The grains differ from each other in the arrangement of unit cells, although the type of lattice in all of the grains is the same. Within a single grain, the arrangement of the crystalline unit cells is the same (Fig. 4).

Characteristic of real metals is the disorderly arrangement of atoms at a number of places. This is particularly true of atoms

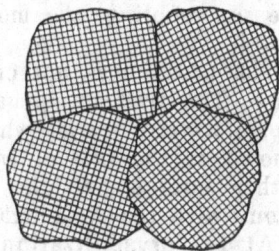

Fig. 4. Schematic of metal structure

2*

located at the grain boundaries. If the disposition of the atoms is disturbed, the properties of the metal change both at the grain boundaries and throughout the body of the metal.

Grain boundaries, for example, are more amenable to etching than other regions of the grain. Besides, grain boundaries are generally stronger than the grain proper. Therefore, the strength of a metal can be enhanced by reducing the size of its grains, since the area of the boundaries is much greater in a fine-grained than in a coarse-grained metal.

2.2. Crystallization of Metals

If a solid metal is heated, it melts at a temperature which is characteristic of each metal. The geometrically orderly arrangement of its atoms changes into a chaotic one. The liquid state of a metal is possible only at a temperature above the melting point. The cooling of a molten metal results in its solidification, or crystallization, which begin at a point theoretically equal to the melting point of the solid metal.

On crystallization, the atoms of the metal are grouped again into crystalline cells with an orderly arrangement of atoms, characteristic of the metal in question.

Solidification takes some time to begin after the melting point has been attained, the metal supercooling in the process to a lower temperature. On supercooling, the energy of atoms decreases, which is advantageous from the standpoint of their orderly arrangement in crystal. The greater the supercooling, the more rapid is the solidification of a metal.

The metal does not crystallize in all of the volume at once. The first to appear in the liquid are the so-called *crystallization nuclei* which grow at a predetermined rate known as the *rate of crystal growth*. The nuclei may be either foreign particles (impurities) or stable groups of atoms with the same arrangement as that of the solid metal.

At the crystallization point, the extraction of heat is compensated for by the *latent heat of crystallization*, i.e., the excessive energy of atoms that have already assumed a geometrically orderly arrangement. In some metals, the latent heat of crystallization (or melting) evolves at such a high rate that the temperature of the solidifying metal even rises.

The process of crystallization may be described in terms of two main characteristics, *viz.*, the number of nuclei formed in unit volume per unit time, and the linear growth of the crystal per unit time. Both characteristics are variable, they vary with the supercooling temperature, each in a different way. This means that each solidification temperature is related to a different number of nuclei and a different rate of crystal growth.

The relationship between these values determines the size of the resulting crystals, i.e., the size of primary grains in solid metal, and, therefore, many properties of the metal. When there are few centres and the growth rate is high, the crystals are large, while when the centres of crystallization are numerous and the growth rate is low, a multitude of fine crystals are formed. Therefore, the grain size in a crystallizing metal may be controlled within certain limits through the effect of supercooling on the number of crystallization nuclei and their rate of growth.

The greater the rate of crystallization, the finer the grains and the higher the strength of the casting. On the contrary, large crystals and lower strength of the metal are obtained when the cooling is slow.

The growing crystals have initially a geometrically regular shape, but as they move ceaselessly in the liquid, they impinge upon one another, hinder their growth, and lose their regular external shape. The different faces of the crystal may have unequal rates of growth, this resulting in curvilinear grain boundaries (Fig. 5) of real structures.

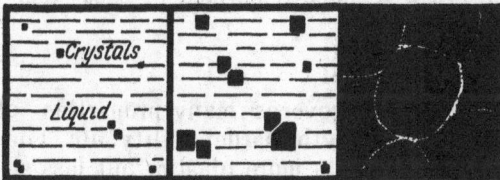

Fig. 5. Formation of crystalline structure on solidification of metal

Metallic crystals may also differ in shape. Characteristic of steels are tree-like crystals—the so-called *dendrites* — with a great number of mutually perpendicular axes branching off the main axis. Dendrite axes contain less impurities than the inter-axis regions. This is due to the fact that crystallization is not instantaneous, but goes on as the metal cools gradually. It is apparent, then, that the first to crystallize will be the microvolumes with a higher point of

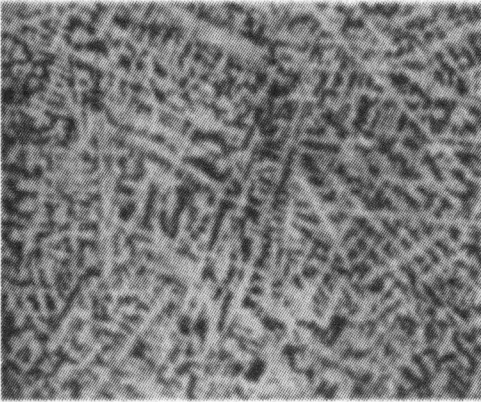

Fig. 6. Dendritic structure of cast steel

crystallization, this being characteristic of the microvolumes the least contaminated with impurities. After the dendrites have been formed, the remaining portion of the melt to which the impurities are driven, crystallizes in the inter-dendritic regions.

Therefore, a characteristic dendritic pattern, composed of bright axes and darker inter-dendritic regions (Fig. 6), is observed when a smooth surface of cast steel is etched by an acid.

2.3. Grain Size in Steel

The size of grain governs many properties of metals, such as plasticity, strength, hardenability, etc. For example, a coarse-grained metal is more plastic, but less strong than a fine-grained one, and a higher strength may be imparted to it by reducing the grain size.

As has been mentioned, the structure (grain size included) is dependent upon both the composition of the steel or alloy and the treatment it has received. Therefore, grain size may be adjusted by appropriate changes in composition, by plastic working, or by heat treatment. Consider these methods in greater detail.

Pure metals generally form coarse grains; this is also true for many alloys, e.g., stainless steels. Small-size castings sometimes show columnar crystallization with elongated crystals stretching out from the surface to the centre of the casting. The mechanical strength of such a casting is low.

A finer structure in steels or alloys may be obtained with the aid of special additives termed *inoculants* and this technique for altering the structure of the metal is referred to as *inoculation* (or modification). Inoculation is essentially the introduction into a molten metal of certain substances, such as aluminium, boron, titanium, zirconium, and others.

The modifying action of the additives may have a dual effect.

1. The inoculant reacts with the impurities and gases to give fine particles of refractory compounds, which serve as additional nuclei of crystallization.

2. Many additives are surface-active substances with respect to the base metal, i.e., they concentrate at the surface of the growing crystals and thus hinder their further growth.

In either case, the grains of the modified steel become finer, and steel strength increases. If steel is heated above 900-950 °C, its grains coarsen and the structure remains coarse-grained after cooling to room temperature. In such cases, modification is also beneficial as it retards the grain growth in steel to higher temperatures. In consequence, a steel doped with elements which reduce the size of grains is more amenable to heat treatment, as it is less likely to be affected by overheating.

Grains of steel are very sensitive to plastic deformation. Loading of a metal may produce an elastic deformation which disappears once the load is removed. Steel grains in this case experience no change whatsoever (Fig. 7a). In plastic deformation, on the contrary, the shape and size of grains change, as does the overall configuration of the metal involved.

In the course of deformation, one part of a grain moves with relation to another along a predetermined plane, termed *slip plane*. In addition, the displaced portions of grains tend to orient their slip planes along the direction of force application. Besides, the grains split into still finer particles, this resulting in a fine-grained structure. When the deformation ratio is considerable and the direction steady (as in rolling), the grains get oriented along this direction. Such a structure is called *texture* (Fig. 7b).

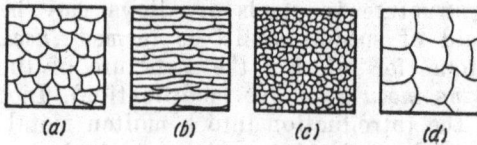

(a)　　(b)　　(c)　　(d)

Fig. 7. Structural changes

(a) initial structure; (b) after cold working; (c) after heating, first stage of re-crystallization; (d) second stage of recrystallization

Splitting of grains in the course of plastic deformation increases hardness and strength, but impairs toughness. Such a strengthening of metal is called *work hardening*.

In many a case, work hardening is objectionable as it hinders subsequent machining of the steel. At the same time, plastic deformation is essential for forcing steel into the required shape and size (as by pressworking, wire drawing, etc.).

Therefore, it is necessary to eliminate the effect of work hardening and enhance the plasticity of the metal. This is achieved by heating the cold-hardened metal to a predetermined temperature. Fine grains, resulting from a deformation, are unstable because of a large boundary surface close to which atoms are disturbed from the orderly arrangement characteristic of the metal involved. Therefore, heating, which communicates additional energy and mobility to atoms, is certain to facilitate their rearrangement in accordance with the type of crystalline cell.

The first effect of heating is that the lattice deformations are eliminated, the process being called *recovery*. Plasticity is improved and hardness decreases, but no change in the grain structure of the metal occurs and the properties of the metal remain high. Therefore, heating to a much higher temperature is required to coarsen the grains and restore the original structure and properties.

The growth of grains and attendant phenomena which take place in the course of heating of a plastically-deformed metal are called *recrystallization*.

Recrystallization proceeds (after the recovery) in two stages:

1. Primary recrystallization which consists in the formation of fine randomly-oriented grains instead of those oriented directionally by deformation (Fig. 7c).

2. Secondary recrystallization consisting in the blending of fine grains into larger ones.

The higher the temperature and the longer the holding time, the coarser the recrystallized grains; the size of the latter may exceed that of the initial (prior to deformation) grains (Fig. 7d). Recrystallization temperature of commercially pure metals is approximately equal to 0.4 the melting point of the metal in question.

It can easily be calculated that the iron recrystallization temperature is approximately 450 °C. In less pure metals and in alloys, the recrystallization temperature is higher.

It follows from the above that if a metal is deformed at a temperature above the recrystallization point (hot working), neither cold working nor strain-hardening occurs. If a metal is deformed below the recrystallization point (cold working), heating is necessary to restore its original properties.

Thus, the structure of a metal depends on its chemical composition and the working it is subjected to.

2.4. Methods for Investigating the Structure of Metals

Investigation of the structure of a metal is often necessary in order to find its quality after working, to find the causes of a failure, etc. A distinction is made between *macroscopic* and *microscopic structures*.

Macroscopic structures are studied by a naked eye or at low magnification ($\times 500$). This allows to determine the continuity of the metal; presence of cracks, pores, or cavities; dendrite structure; chemical heterogeneity; structure and grain size in fracture, etc.

Macrostructure investigations require special preparation which consists in machine or hand polishing of the surface under test. The prepared surface is then etched by various reagents. For example, hydrochloric acid or a mixture of hydrochloric and sulfuric acids are the etchants most frequently used to bring out the dendrite structure (see Fig.6). Carbon and many other elements are distributed unevenly in microvolumes of steel, therefore, different structural regions submit to the etching unequally. For example, the axes of dendrites contain less impurities, and are less amenable to etching.

Widely used is the sulfur-print method. Photographic paper moistened with sulfuric acid is applied against the surface of a well-polished steel specimen and held several minutes. Sulfur in steel is in the form of fine particles of iron and manganese sulfides which react with the sulfuric acid. The evolving hydrogen sulfide and the photographic emulsion produce a silver sulfide, with the effect that areas of the paper in contac with the surface sulfides grow darker and reveal accumulations of sulfur.

Microscopic structure is studied with the aid of special microscopes at magnifications of about 2,000 times. Grains and other structural items are much smaller in size than the surface irregularities remaining after polishing. Therefore, a specimen intended for a microstructural investigation (a microsection) should be thoroughly polished and burnished until an absolutely even surface is obtained. The section is then etched with various reagents; the etching compound generally used for carbon steels is a 2-5 per cent solutuion of nitric acid in ethyl alcohol.

Studies fo microstructure yield information on the distribution, shape, and size of structural constituents, size of grains, quantity of harmful non-metallic inclusions, etc.

CHAPTER 3

Equilibrium Diagram of Iron-Carbon Alloys

Pure iron is very seldom used in engineering applications as it lacks in many necessary properties. The chief construction materials in engineering and construction are steels

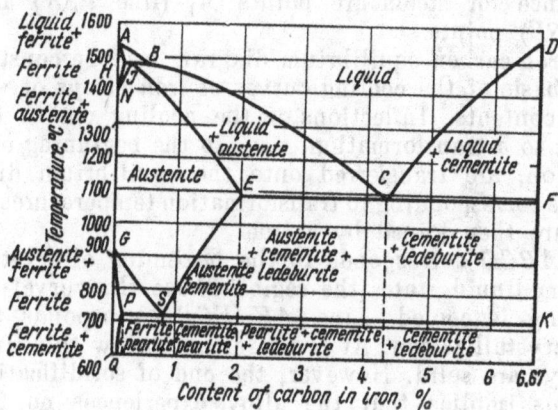

Fig. 8. Iron-carbon equilibrium diagram

and cast irons, i.e., alloys of iron with carbon. It was observed that addition of a small amount of carbon to iron substantially affects its strength and toughness. Besides, steels and cast irons are heat-treatable and thereby their properties can be changed to a still greater degree.

As has been mentioned above, the phase of iron and carbon solubility in it differ at different temperatures. Therefore, iron-carbon alloys may also occur in various phases at various temperatures.

The changes of the phase, constitution and, in consequence, properties of steels and cast irons in the process of cooling are described by the iron-carbon equilibrium diagram shown in Fig. 8.

Plotted on the horizontal axis of the diagram is the quantity of carbon alloyed to the iron. Maximum carbon content of 6.67 per cent is found in cementite. Iron alloys with high carbon contents have been poorly studied and are not used in practice.

Temperatures are plotted on the vertical axis. The lines on the equilibrium diagram limit the regions of stable state of various phases; they show the temperature and direction of changes in the constitution of a given iron-carbon alloy.

These temperatures are called *critical points*. Of major importance for steels are points A_1 (line *PSK*) and A_3 (line *GSE*) points.

The iron-carbon equilibrium diagram may be constructed on the basis of the cooling curves of iron alloys of various carbon contents. Inflections on the cooling curves corresponding to a transformation, e.g., to the beginning of crystallization, are transferred onto the equilibrium diagram as points corresponding to transformation temperatures; these points are then joined by a line.

Line *ABCD* corresponds to the beginning of crystallization from liquid state (the region above the curve) as the iron alloy is cooled. Line *AHJECF* corresponds to the end of crystallization. At temperatures below this line, all the alloys are solid. However, the end of solidification by no means implies that the alloy experiences no further transformations. Major compositional changes may occur in a solid metal. This is due to rearrange ments of the iron crystal lattice (see Para 2.1) and to the various solubility of carbon in irons of different modifications.

Structural constituents of steels and cast irons are generally designated as follows:

Cementite is a chemical compound of iron and carbon, Fe_3C. It is very brittle and hard (which explains its name).

Ferrite—the solid solution of carbon (up to 0.02 per cent) in alpha-iron—is a very plastic structural constituent. As was already mentioned, alpha-iron has a body-centred cubic lattice; delta-iron, which exists above 1,400° C, has the same type of lattice, but of a larger constant. Thus, there are a high-temperature ferrite existing in region *AHN* and a low-temperature ferrite found in region *GPQ*. The word "ferrite" is derived from the Latin "ferrum"—iron.

Austenite is the solid solution of carbon (up to 2 per cent) in gamma-iron. Similarly to pure gamma-iron, austenite exists in carbon alloys only at high temperatures (above 723°C). The plasticity of austenite (so named after the famous investigator Roberts-Austen) allows this type of steel to be forged and rolled.

Pearlite is a structure composed of alternating regions of ferrite and cementite. Ferrite and cementite may coexist in steel as separate aggregates (when formed at different temperatures) or as pearlite (when formed simultaneously at 723°C). Pearlite, which is a mechanical mixture of ferrite and cementite, has higher hardness and lower plasticity than ferrite (the name pearlite is due to its pearly lustre).

Ledeburite is a mixture of pearlite and cementite, characteristic of white cast irons. It is named after the famous German metallurgist Ledebur.

Ferrite, pearlite, and cementite, taken in various proportions, form the structure of a carbon steel. The actual properties of steel depend on the quantities of these constituents. For example, if a steel consists of ferrite only (almost pure iron), it is soft and ductile. The greater the amount of carbon in a steel, the higher is the content of pearlite, the higher is its hardness, and the lower its ductility. Finally, as the carbon content in steel exceeds 0.8 per cent (no ferrite is present), cementite appears which increases the hardness and brittleness of the steel.

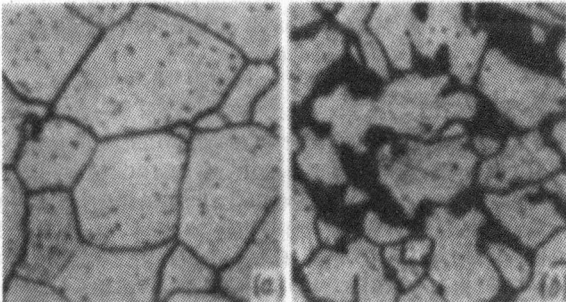

Fig. 9. Microstructure of steels (×100)

(a) pure iron, ferrite grains; (b) 0.2% C, ferrite (bright grains) and pearlite (dark grains)

Figure 9 shows typical structures of carbon steels. At carbon content below 0.02 per cent, steel structure consists of ferrite only (Fig. 9a). When carbon ranges from 0.02 to 0.80 per cent, the steel has a double-phase structure composed of ferrite and pearlite (Fig. 9b). Steel with 0.80%C has a pearlite structure. Steels of a higher carbon content (up to 2.0 per cent) contain pearlite, but also cementite in the form of a grid along the boundaries of grains.

Illustrated in Fig. 10a is a cast-iron structure composed of pearlite and cementite. Cementite in cast iron is unstable; during annealing, it decomposes giving off free carbon in the form of graphite; the structure of the cast iron will then not conform to the equilibrium diagram, shown in Fig. 8, and will consist, for example, of pearlite, ferrite, and graphite (Fig. 10b).

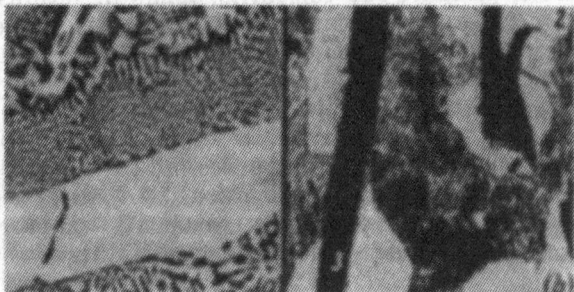

Fig. 10. Microstructure of cast irons (×100)

(a) white iron; (b) grey iron
1 — pearlite; 2 — ferrite; 3 — graphite

Graphite is a very weak structural constituent, it fails readily, which makes the cast iron stiff and brittle. Structural graphite may also be obtained in casting by adjusting the cooling conditions or by introducing special additives.

Graphitic cast iron has a dark fracture and is called grey in contrast to white iron whose fracture is bright. White iron is even more brittle and hard as its structure includes much cementite.

From all the above said it follows that the iron-carbon equilibrium diagram is very helpful in investigating the changes which occur in ferrous metals in the process of heating or cooling.

It was the eminent Russian metallurgist D. Chernov (1839-1921) who suggested that a graphical method should be used for linking structural changes with temperature. D. Chernov also formulated a theory of steel crystallization, described the structure of a steel ingot, and put forward a theory of heat-treatment processes.

CHAPTER 4

Physical and Working Properties of Metals

4.1. Physical Properties of Metals

The structure of metals is interrelated with their physical properties, and the properties of alloys differ from those of their starting materials. Many physical properties depend on structural changes occurring in heat treatment, cold working, and other kinds of processing.

Density. Density is mass of a substance per unit volume. Most metals have a high density (Table 1). In recent deca-

Table 1

Physical Properties of Metals

Metal	Atomic symbol	Density at 20°C, g/cm³	Melting point, °C	Electric resistivity, 10^{-6} ohms cm	Heat conductivity at 20°C, W/m degree	Thermal expansion factor at 20°C, 10^{-6}/degree
Iron	Fe	7.86	1,539	10 00	83.8	11 50
Aluminium	Al	2.70	660	2.62	143.5	23.10
Manganese	Mn	7.46	1,244	71.00	5.02	22.10
Copper	Cu	8.92	1,083	1.55	1,650 0	16 50
Molybdenum	Mo	10.30	2,625	5.17	145 0	495.60
Nickel	Ni	8.90	1,455	7 24	82 9	13.50
Chromium	Cr	7.14	1,875	12 90	67.0	6 20
Tungsten	W	9 63	3,410	5.03	201 0	434.50
Titanium	Ti	4.50	1,723	47.50	15.1	7.14
Lead	Pb	11.34	327.4	21.90	35.6	28.10
Zinc	Zn	7.14	419.5	6.10	112.0	32.50
Gold	Au	19.30	1,063	2 06	1,240 0	14 20
Vanadium	V	5.96	1,700	26.00	31.0	8.30
Tin	Sn	7.29	232	11.40	65.7	46.60
Cobalt	Co	8.71	1,480	5.03	71.1	12 50

des, light metals (titanium, beryllium, magnesium, aluminium) and their alloys are finding an ever increasing application, particularly in aerospace industries which make use
of their high strength-to-weight ratio. When manufacturing
alloys, it is necessary to take into account the difference in
densities of starting materials in order to obtain homogeneous alloys.

Melting point. The temperature at which a metal passes
from solid to liquid state is called its melting point (Table 1). It characterizes the cohesive force between the atoms
of the metal.

The melting point of an alloy may be either higher or
lower than that of the starting substances. It can be seen
from the iron-carbon equilibrium diagram (Fig. 8) that
melting point of iron is equal to 1,539°C, while that of
cementite is about 1,600°C. At the same time an iron alloy
with 4.3% C has a melting point of a mere 1,130°C. By
contrast, aluminium and nickel, when taken in a predetermined proportion, form an alloy with a higher melting
point than that of either metal.

The melting point of any alloy may be found from its
equilibrium diagram, which shows the melting points of
all alloys of the given components. For example, the iron-
carbon equilibrium diagram may serve to determine the
melting points of all carbon steels and cast irons.

Heat conductivity. The quantity of heat, watts, passing
per unit time across a slab 1 cm² in cross section by 1 cm
thick when the temperatures on the faces of the slab differ
by 1°C, is termed *heat conductivity*. Metals have a higher
heat conductivity than other substances, with copper, silver,
and gold being foremost in this respect. Heat conductivity
of a metal or alloy is temperature-dependent. Table 1 lists
values of heat conductivities of various metals at 20°C.

Heat conductivity should be taken into account when assessing the behaviour of a part in service. For example, parts
operating at high temperatures (valves, pistons, etc.) should
be manufactured from alloys or metals of high heat conductivity in order to avoid loss of mechanical strength.

Electrical resistivity. The electrical resistance of a metallic conductor 1 m long by 1 mm² in cross section is named its *resistivity*. It can be seen from Table 1 that the

electrical resistivity of metals varies within wide limits. Metals with the least resistivity (copper, aluminium) are used to manufacture electrical conductors. Metals and alloys with a high resistivity are used in various devices which convert electrical energy into heat. For example, nichrome — an alloy of nickel and chromium—is used to manufacture spiral windings of electric heaters, electric irons, etc.

The various structural constituents (ferrite, cementite) have different resistivities which affects the overall resistivity of a ferrous metal.

Thermal expansion coefficient. This parameter characterizes the variation in the length of a metallic specimen as it is heated by 1 °C, referred to the original length. The thermal expansion coefficient is also temperature-dependent.

The tendency of metals to expand when heated should be taken into account in most of their applications. A casting with sudden transitions from large to small sections may crack in the process of cooling, because a great difference in cooling rates will give rise to internal stresses exceeding the strength of the metal at transition points. A similar phenomenon takes place in welding of various structures. The thermal expansion coefficient should be given due consideration when a metal operates in contact with some other substance.

Metals with different thermal expansion coefficients are used to manufacture bimetallic thermal relays. Platinite, an iron alloy with 48% Ni, has a thermal expansion coefficient equal to that of common glass and is employed for leads through the glass bulbs of radio tubes.

Magnetic properties. The magnetic induction, retained by a specimen after a magnetic field is removed, is called *remanence*. The field strength required to demagnetize the specimen is called *coercive force*.

An important property of metals is their *magnetic permeability*, i.e., the capacity to concentrate magnetic lines of force. It is equal to the ratio of the remanence (in gausses or teslas) to the coercive force (in oersteds or amperes).

For most metals, the magnetic permeability is close to unity. Metals with a magnetic permeability slightly exceeding unity are called *paramagnetic*, those with the magnetic permeability less than unity, *diamagnetic*.

Metals with a very high magnetic permeability (up to hundreds of thousands of gauss/oersted) are called *ferromagnetic* materials (iron, nickel, cobalt).

Magnetic alloys with small values of coercive force are called magnetically soft. They are employed to manufacture parts subjected to periodic magnetizing and demagnetizing (cores of electric magnets and transformers). The magnetically hard alloys with a high coercive force are used for permanent magnets.

Magnetic properties, as many other physical characteristics, vary with the structural state of alloys and may serve as a means for investigating the transformations in alloys.

4.2. Working Properties of Metals

Suitability of a metal for the manufacture of a structure or part cannot be always expressed in terms of physical and mechanical properties. The manufacturing or service conditions of a part are usually more severe than those of the tests. A more accurate assessment of the quality of a steel is obtained by means of the so-called working tests.

Forgeability test. Steels of the same grade, but from different heats, may differ in plasticity, and therefore a preassessment of plasticity is required before a hot rolling procedure is chosen.

Forgeability tests are conducted on specimens of a mass up to 1 kg, cast in the process of a heat or teeming. Cylindrical specimens are forged to a square bar 15×15 mm. The bar is then bent through 180° by means of a power hammer until its sides touch each other.

Forgeability is considered good if no tears, cracks, etc., appear. If the tears are small, forgeability is assumed to be satisfactory. Large tears are evidence of poor hot-rolling properties of the steel under test. A metal can be rolled only if it possesses a good forgeability, since rolling requires greater plasticity than forging.

Bending test for forge weldability. Welding tests serve to determine the ability of a steel to bend at a forge-welded point in a specified manner and within specified limits.

A specimen is cut out of the metal under test, its size being a function of the profile and the thickness of the

metal. The specimen is then cut transversely in the middle
into two parts. The ends of the cut are upset and pressure-
welded together at an angle of 30-45° to the longitudinal
axis of the specimen by lap welding (Fig. 11a).

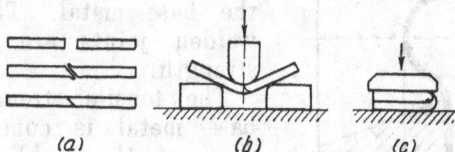

Fig. 11. Bending test for forge weldability

The test consists in bending the specimen at the weld
in accordance with one of the following methods: bending
to a specified angle (Fig. 11b); bending around a mandrel
until the sides are parallel; and bending until the sides
come to contact (Fig. 11c). The bending method should be
indicated in the specifications.

Absence of cracks, tears, peelings, or fractures after
bending is evidence that the steel specimen has withstood
the test.

Sheet steel is frequently tested for bending by the above
method, but without cutting or welding of the specimen.
Preservation of continuity throughout the test indicates
that the specimen has stood up to the test.

Test for deep drawing. A qualitative characteristic of the
capacity of a thin sheet metal to be cold-stamped is obtained by subjecting the metal to a test for deep drawing.

A strip of sheet steel 70-90 mm wide is clamped in a
die of a special apparatus. The die has a round opening.
A cavity is drawn in the specimen (Fig. 12) with the aid
of a spherical-point punch. The test is conducted until
cracks and tears appear on the convex portion, this being
ascertained with the aid of a mirror. The drawing capacity
of the metal is indicated by the depth of the cavity which
can be pressed before the cracks appear; the punch diameter
is chosen to suit the thickness of the metal under test.
Materials other than steels, e.g., brasses, are also tested
for deep drawing.

3*

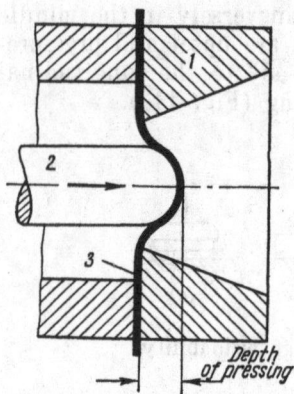

Fig. 12. Testing of sheet metal for deep drawing

1 — die; *2* — punch; *3* — specimen

Mechanical testing of welded joints. The properties of a metal in proximity to a weld and those of the weld may differ greatly from the properties of the base metal. This is why welded joints are tested for strength.

The tensile strength of the base metal is compared with that of the weld by testing specimens shown in Fig. 13. The weld seam is machined flush with the base metal. The tensile strength is determined in the same manner as for solid specimens, reporting the place of failure (base metal, weld, heat-affected zone), the presence of flaws in the fracture (cleavage, lack of fusion, etc.).

Flat specimens with the weld seam removed flush (Fig. 14) may also be used for tensile tests. The ultimate strength is determined as the ratio of the force at failure to the cross-section area multiplied by a factor which, for the carbon and low-alloy steels, is equal to 0.9.

When the strength of a weld is known to be smaller than that of the base metal, plain (not notched) specimens are used, and in calculating the ultimate strength the above factor is neglected.

Welded joints may be tested by bending over to a predetermined angle or until the sides are parallel. The set-up (Fig. 15) is similar to that which is used for solid metal. The welded joint is located along or across the mandrel, and crack appearance is recorded. The bending tests give an assessment of the welded joint plasticity.

There are also other working tests for metals, such as jumping-up; bending over and twisting (for wire); flattening (for pipes); reaming and forming of a double scarf joint (for sheet steel).

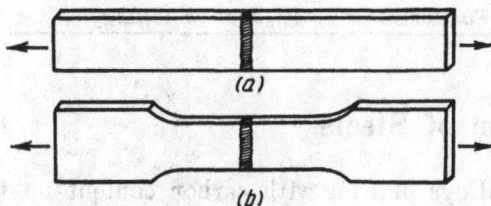

Fig. 13. Flat specimens for tensile testing of welded joints

Fig. 14. Specimen for weld testing in tension

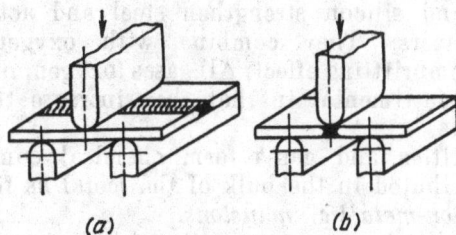

Fig. 15. Bend testing of welded joints

(a) across weld; (b) along weld

CHAPTER 5

Classification of Steels

Steels are alloys of iron with carbon content up to 1.7 per cent. Carbon enhances the strength of steels (see Fig. 8).

Besides carbon, all of the steels contain impurities: silicon, manganese, sulfur, and phosphorus. Phosphorus and sulfur are detrimental impurities which are very difficult to remove during steel manufacture.

Phosphorus embrittles steel, particularly at low temperatures (*cold shortness*); its admissible content is usually within 0.03-0.04 per cent. Sulfur causes brittleness of steel at higher temperatures (*red shortness*) as it forms low-melting compounds with iron. Sulfur content in finished steel is limited to 0.03-0.04 per cent.

The harmful effect of sulfur may be countered to a great degree by introducing manganese, this giving rise to compounds of sulfur and manganese with a higher melting point and strength than have the iron-sulfur compounds.

Manganese and silicon strengthen steel and act as effective deoxidizers. They combine with oxygen, thus restraining its embrittling effect. All gases (oxygen, nitrogen, hydrogen) are detrimental in that they increase the brittleness of steels.

Thus, impurities and gases form chemical compounds which are distributed in the bulk of the metal as fine particles called *non-metallic inclusions*.

On the other hand, manganese, silicon, and nitrogen may serve as alloying elements intentionally introduced into steel to impart the desired properties to it.

According to chemical composition, steels may be subdivided into *carbon steels*, containing carbon and impurities, and *alloy steels*, which, additionally, contain one or more alloying elements.

Common grade carbon steels are classified by Soviet State Standard GOST 380-60 into two groups and one subgroup.

Steels of group A are specified in terms of mechanical properties and designated by letters Cr. and digits 1, 2, etc. (see Table 2).

Table 2

Properties of Some of Group A Steels

Steel	Ultimate strength, MN/m²	Minimum yield point, MN/m²	Minimum per cent elongation	Cold bending through 180°: s=specimen thickness; d=mandrel diameter
Ст. 0	314	—	22	$d = 2s$
Ст. 1 Ст. 1 кп	314-392	—	33	$d = 0$
Ст. 3 Ст. 3 кп	373-461	206-235	27-25	$d = 0.5s$
Ст. 5	490-609	225-275	21-19	$d = 3s$
Ст. 7	686-735	—	10-11	—

Table 3

Chemical Composition of Some of Group Б Steels

Grade of steel	Content, per cent				
	carbon	silicon	manganese	phosphorus, max	sulfur, max
МСт.0	0.23 max	—	—	0.070	0.060
МСт.3 кп	0.14 - 0.22	0.07 max	0.30 - 0.60	0.045	0.055
МСт.3	0.14 - 0.22	0.12 - 0.30	0.40 - 0.65	0.045	0.055
БСт.3 кп	0.12 max	0.07 max	0.25 - 0.55	0.080	0.060
БСт.3	0.12 max	0.12 - 0.35	0.25 - 0.55	0.080	0.060

Group Б steel, specified in terms of chemical composition, is additionally coded by letters М, Б, К placed before the grade designation and indicating the method of manufacture (М for open-hearth, К for oxygen converter, Table 3). Steel of subgroup В is supplied on the basis of mechanical properties and chemical composition. Rimmed steels are additionally designated by letters "кп", semi-killed steels, by letters "пс".

High-grade structural carbon steels to GOST 1050-60 must meet more stringent requirements to chemical composition, chiefly with respect to impurities. These steels are designated by two-digit numbers indicating the mean content of carbon in hundredths of a per cent: 05кп, 08кп, 10кп, 10, 20кп, 20, 25, 30, 35, 40, 45, etc. The letter Г following the designation of steel grade (14Г, 18Г) points to an elevated manganese content.

Common and high-grade carbon steels in the form of rolled sections, plates, and sheets are used in mechanical and construction engineering, but they cannot provide the required variety of properties.

Automatic steels with a higher-than-common sulfur and phosphorus content are employed for machining non-critical parts. The elevated sulfur and phosphorus content ensure free-machining properties, better surface finish, and higher efficiency of machine tool. The steel designation denotes the mean carbon content in hundredths of a per cent, preceded by letter А, e.g., А12, А30.

Carbon tool steels used for the manufacture of cutting tools are designated by a letter У and the mean quantity of carbon in tenths of a per cent (e.g., У8, У10).

Letter А which follows the steel grade designation indicates low sulfur and phosphorus content.

All steels, irrespective of application, should comply with State Standards as to chemical composition, properties, and working.

Alloy steels are coded with letters indicating the alloying elements followed by digits which specify the per cent content of the element. The amount of carbon in hundredths of a per cent is given by the initial digits. Absence of digits implies that the quantity of the element is below one per cent.

Alloying elements are conventionally designated by the following letters:

A—nitrogen	П—phosphorus
Б—niobium	P—boron
В—tungsten	C—silicon
Г—manganese	T—titanium
Д—copper	Ф—vanadium
К—cobalt	X—chromium
M—molybdenum	П—zirconium
H—nickel	Ю—aluminium

For instance, a steel designated 60C2ХФА has the following composition: 0.56-0.64% C, 1.4-1.8% Si, 0.9-1.2% Cr, 0.10-0.20% V, and a very low amount of sulfur and phosphorus.

By application, carbon and alloy steels fall into the groups below:

1. Structural steels used for the manufacture of machine elements. Most of these steels are heat-treated either after final machining or, sometimes, in the process of metal manufacture at an iron and steel plant (construction steels, low-alloy steels).

2. Tool steels intended for the manufacture of cutting tools, dies, etc.; high-speed steel belongs to this category.

3. Special steels, e.g., stainless, wear-resistant, temperature-resistant, heat-resistant, acid-resistant, those with special electric or magnetic properties, etc.

REVIEW QUESTIONS

1. Describe the types of mechanical tests differing in the rate of load application to the specimen.

2. What are *yield point* and *ultimate strength*? How are they determined?

3. How is Brinell and Rockwell hardness determined?

4. Name the types of simple crystalline lattices. In what respect does a real metal differ from an ideal crystal?

5. What changes take place in a metal in the process of solidification?

6. Name the basic methods of refining the grains of an alloy.

7. What changes occur in a grain in the process of plastic deformation and subsequent heating?

8. What is *recrystallization*?

9. Name the structural constituents of steels and cast irons and speak about their basic properties.

10. How are microstructure and macrostructure of a metal investigated?

11. What are the main physical properties of metals?

12. What is included in the tests for forgeability, forge weldability, and deep drawing?

13. In what manner are welded joints tested for strength?

14. Speak on steel classification by chemical composition and application.

15. Explain classification and designation of carbon steels.

Iron is obtained by reducing iron oxides into metallic iron in blast furnaces. The conditions in these furnaces are such that iron is saturated with carbon. In addition to carbon, the pig iron contains silicon, manganese, sulfur, and other impurities.

Blast furnace is the longest known unit for iron making. Ancient blast furnaces, bloomeries, were built from clay. They were small in size and produced a few kilograms of iron per day. Gradually blast furnaces grew in size. Today's blast furnaces are gigantic structures made of steel and refractories. They are capable of producing up to 6,000 tons of pig iron every 24 hours.

CHAPTER 6

Starting Materials for Blast Furnace Smelting

Iron is produced in blast furnaces by smelting iron and manganese ores. The fuel is coke, fuel oil, natural gas, or pulverized coal. Fluxes and metallic additives, such as scale, hearth cinder, and metallic scrap are also charged into blast furnaces.

6.1. Iron Ores

Iron ore is a mineral occurring naturally and containing iron oxides. The mineral oxides are accompanied by gangue, i.e., earth which carries no iron. The greater the amount of gangue in an iron ore, the more difficult and the costlier the recovery of iron. Used for blast furnace smelting are those minerals which ensure an economically advantageous extraction of iron. These minerals are what is called ore.

Classification of iron ores by mineralogical composition. Iron may be found in iron ores in the form of ferric oxide Fe_2O_3, magnetic oxide Fe_3O_4, and/or ferrous oxide FeO. The latter never occurs singly, but is commonly combined with carbon dioxide forming iron carbonate $FeCO_3$.

Ferric oxide Fe_2O_3 may exist either separately or in combination with water molecules. Anhydrous ferric oxide is called *hematite* or *red iron ore*. The gangue of this ore is chiefly *silica* SiO_2. Hydrous ferric oxide is *limonite* or *brown iron ore*; its gangue is composed of *alumina* Al_2O_3, silica SiO_2 and *lime* CaO.

Magnetic iron oxide Fe_3O_4 is called *magnetite* or *black iron ore*. The gangue of this type of ores may contain silica, lime, and *magnesia* MgO.

Iron carbonate $FeCO_3$ is known as *siderite* or *sparry iron ore*.

Pure (free from gangue) hematite and magnetite contain respectively 70 and 72.4 per cent iron. However, actual iron content in these iron ores is less because of the gangue.

6.2. Preparation of Ores for Smelting

An as-mined ore seldom possesses the properties required for blast furnace smelting. Smelting of ores composed of very fine particles or, by contrast, of very large lumps would be a very poor practice indeed. Furnace performance is also impaired when the ores are high in gangue and low in iron (the so-called *lean ores*). In many a case, ores contain harmful impurities which are to be removed (be it only partially) prior to smelting the ore in a blast furnace. *Concentration of ores* aims at imparting the required properties to ores.

Crushing. Blast furnaces should be charged with ore lumps not larger than 100 mm. The run-of-mine ore consists mostly of much larger blocks which require crushing. Ores subject to concentration treatment also require crushing.

Ores may be crushed and ground in a number of ways. In some instances, ores are comminuted by crushing or splitting, in others, by impact or grinding. This purpose is served by *crushers*.

If an ore is rich in iron and requires no concentration, it may be charged directly into blast furnaces after crushing.

Screening. Screening consists in passing the material through sieves of special screens. Finer lumps falling through the sieves are called the *undersize*. The larger lumps (the *oversize*) remain on the screen. In this manner ore is sorted into several size classes.

Grinding and sizing. A lean ore requires *concentration*, i.e., reduction of its gangue content. The effect of gangue removal is a higher iron content in the ore. However, the grains of gangue and iron oxides are intimately mixed. Removal of the gangue requires the separation of these grains, this being achieved by milling the ore to a fine powder.

The ore is ground in special mills. Generally, ores are ground with the admixture of water (*wet grinding*). Water mixed with finely ground ore forms a creamy mass, or the *pulp*. Besides fine ore particles, the pulp contains larger ones which have failed to get crushed in the mill. These particles of sand size require recycling, for which purpose they must be separated from the fine ones. To do so, the pulp is directed into special classifiers (Fig.16). Pulp is

Fig. 16. Bowl-rake classifier

admitted into the classifier trough via a narrow launder. The velocity of the pulp in the trough is so small that the larger particles settle out at the bottom. The finer particles remain suspended in the liquid which carries them out of the trough and to the concentrators. The larger particles, called *sand*, settle down in the trough, then are reclaimed by the rake and vanes to be recycled in grinding devices.

Concentration. In the process of concentration, the particles of gangue are separated from those of iron oxides. There exist a number of concentration techniques. The simplest is hand sorting. Lumps of gangue differ from the ore in colour and are easy to separate manually.

Other techniques are based on the fact that gangue and ore differ in density. One of these techniques is washing. A jet of water washes off and carries away the lighter gangue particles, while the ore particles are left over.

Extensive application has been found by electromagnetic concentration. Unlike gangue particles, many iron ores are magnetic. This property of the ores is put to use in a magnetic separator (Fig. 17). A fixed electromagnet is located inside a revolving copper drum, and ore is fed upon the drum from a conveyor. The magnetic ore particles are retained on the drum until they pass the magnet and fall into the concentrate bin. Gangue particles are not attracted by the electric magnet and are, then and there, dumped into the tailings bin.

Next, the ore concentrate is dried. However, this powdery material is not suitable for blast furnace smelting; it must first be agglomerated into lumps.

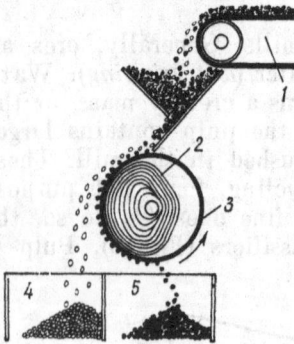

Fig. 17. Magnetic separator

1 — conveyor; *2* — electromagnet;
3 — copper drum; *4* — tailings;
5 — concentrate

Agglomeration. Two methods are generally used: *pelletizing* and *sintering*. A moist concentrate is transformed into lumps (*pellets*) in special drums. In order to make the pellets hard and prevent their disintegration in the blast furnace, they are baked on grates or in special furnaces.

In the process of sintering, air is sucked through a bed of a burden specially prepared of the concentrate, limestone, and fine coke as fuel. The burden is spread uniformly

over the *sintering pallets 3*, which are interconnected carriages with grates (Fig. 18), forming an endless chain. The
chain is passed over special wheels, or *sprockets 4*, which
revolve and drive the chain much as the caterpillar tracks
of a tractor. Located underneath the pallet chain are *suction
chambers*. A vacuum is created in these suction chambers
by means of high-power fans, causing the air to be sucked
through the bed of burden. An igniter *2*, provided at the
point where the burden is charged, kindles the coke in the
burden as it passes underneath it. The bed of burden then

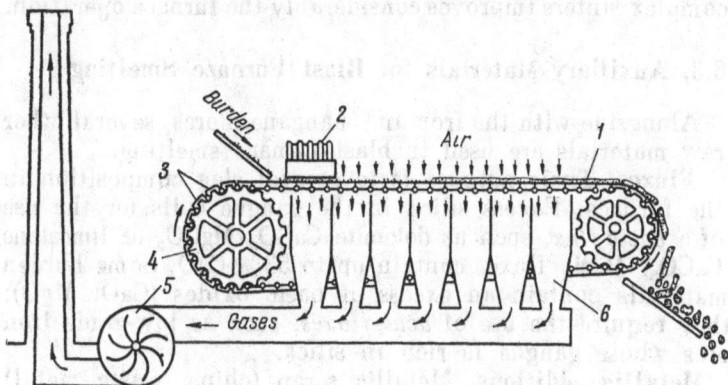

Fig. 18. Sintering machine

1— sinter cake; *2* — igniter; *3* — pallets; *4* — sprocket; *5* — suction fan (ex
hauster); *6* — suction chamber

slides from under the igniter with the coke burning as air
is continuously sucked through the bed. Gradually the bed
is sintered from top to the bottom, the burden lumps
melting on the outside and fusing together to form a
strong product, the *sinter*. At the discharge end, the sinter
is dumped into railway cars or coolers, and then conveyed
to blast furnaces.

In the sintering process, the particles not only fuse together, but some amounts of harmful impurities (sulfur,
arsenic) are removed and necessary additives introduced
into the burden. When some limestone is added to the burden, the product is known as *fluxed sinter*. Sometimes the
full quantity of limestone necessary for blast furnace smel-

ting is added in the process of sintering, the sinter then being termed a *self-fluxing sinter*.

Manganese ore, which is a necessary addition to blast furnace charge, is generally composed of very fine particles. Furnace performance is impaired when it operates on fine ore, and this is why it is good practice to process this kind of ore together with the rest of the burden in sintering machines. A sinter which contains several useful additives is termed a *complex sinter*.

The use of the fluxed and, particularly, self-fluxing and complex sinters improves considerably the furnace operation.

6.3. Auxiliary Materials for Blast Furnace Smelting

Alongside with the iron and manganese ores, several other raw materials are used in blast furnace smelting.

Fluxes. Their purpose is to control slag composition in the furnace. Excess silica in the gangue calls for the use of a *basic flux*, such as dolomite $CaCO_3 \cdot MgCO_3$ or limestone $CaCO_3$. Basic fluxes contain up to 53% CaO. Some burden materials contain an excess of basic oxides (CaO, MgO); they require the use of *acid fluxes*, such as low-grade iron ores whose gangue is rich in silica.

Metallic additions. Metallic scrap (chips, waste metal) is smelted in blast furnaces. Generally, used in blast furnace practice are metallic additions contaminated with clay, sand, broken bricks. Pure metallic additions are used in steelmaking processes as this is more advantageous.

Hearth cinder, scale, slags. Many other iron-bearing materials are used in blast furnace smelting along with the iron ores.

Hearth cinder, running as high as 50 per cent of iron, is the waste product resulting from furnace heating of ingots prior to rolling.

A lot of *scale* is produced when steel is worked in rolling mills; this scale featuring an iron content up to 72 per cent is also fed to blast furnaces.

Open-hearth slag has as much as 25 per cent each of iron and manganese and a considerable amount of calcium oxide, and because of this it is sometimes charged into blast furnaces.

When ferromanganese is smelted, the slag is rich in manganese and may be used for the smelting of some grades of iron.

The stream of gases issuing from a blast furnace carries *flue dust*, or particulate sinter, ore, coke, and limestone. This dust is collected and re-used after sintering.

6.4. Fuel

Formerly, blast furnaces were operated on solid fuel (wood, peat, coal, and charcoal). However, the size of the furnaces grew gradually over the years, and these fuels could no longer meet the stringent requirements placed upon them. Today almost all blast furnaces operate on coke.

Requirements placed on fuel. Solid fuels are subject to the following requirements.

1. *High calorific power.* The quantities of heat, kJ, liberated in the process of combustion of 1 kg of fuel are as follows:

Wood	12,000
Peat	14,700
Brown coal (lignite)	21,000
Coal	34,600
Coke	34,600+

2. *High strength.* Furnace operation deteriorates if fuel disintegrates inside the furnace and gives rise to much fines and dust. This is why fuel lumps should be able to withstand a high pressure from the burden column inside the furnace and show no tendency to decrepitate or pulverize in the process of heating. Only coke is capable of meeting all the above requirements; all other kinds of solid fuels crumble to a varying degree inside a blast furnace.

3. *Minimum content of harmful impurities.* The quality of iron and steel deteriorates sharply if their sulfur content increases. Charcoal contains the least amount of sulfur. Coke has from 0.5 to 2.6% S.

4. *Minimum content of ash, moisture, and volatiles.* Maximum ash content in charcoal is 2.5 per cent, in coke, 11 per cent, in coal, 19 per cent.

Thus, of all the kinds of fuel, coke has the best combination of properties.

Coke. Coke is obtained from coking coals. Before coking, the coal is crushed, washed, and sorted. The various grades are then mixed in predetermined proportions to form a coking charge.

Coking is conducted in special chambers, termed *coke ovens*. The process is essentially dry distillation of the coal mixture heated to 1,000-1,100° C without air access in the course of 11 to 17 hours. A great amount of *coking gas* and volatile matter is released in the process of coking. Coking gas consists chiefly of hydrogen and methane, which have high calorific powers. It is used as fuel in blast and open-hearth furnaces. Volatiles are also collected and processed to obtain valuable chemical products.

In the process of coking, fine coal particles fuse together, a great many pores being formed in the fused mass because of the evolution of the gases. This is why coke is a porous mass very much like a sponge.

When the coking is over, a special pusher forces the coke out of the narrow space of the coke ovens. The hot coked mass disintegrates in the process into elongated lumps 25 to 150 mm long. Red-hot coke is dumped into a cooler car, then transported to a cooling tower where it is quenched with water. Coke ovens are united into groups of 50-60 units, forming a *coke plant*. One such plant produces up to 1,500 tons of coke per day.

Coke differs advantageously in its properties from the coals which go into its making: it is stronger and resists better to decrepitation in blast furnaces, it contains less ash, sulfur and volatile matter, and offers smaller resistance to the passage of gases. Coke is fed to blast furnaces in lumps not smaller than 25 mm. Finer coke finds other applications

The smelting of 1 ton of iron requires from 450 to 1,000 kg of coke, whose cost amounts to 30 to 60 per cent of iron cost. In consequence, the reduction of coke consumption is a major problem facing blast furnace operators. Besides, the reserves of coking coals are running low, and this compels metallurgists to search for new types of blast furnace uels.

Today, natural gas, fuel oil, and pulverized coal are widely used in blast furnace smelting.

Natural gas. Fuel oil. Pulverized coal. The reduction of iron from its oxides is not different from any other reduction process in that it requires a source of heat and a reducing agent. In a blast furnace, both functions are performed by coke. Apparently, an additional amount of reducer, fed to a blast furnace from some other source, will bring coke consumption down.

Such a source may be natural gas which consists chiefly of methane CH_4. As natural gas is injected into a furnace, methane decomposes to form hydrogen H_2 and carbon monoxide CO, both of them being good reducers.

Decomposition of natural gas in a blast furnace requires a certain amount of heat. In order to save the heat in the furnace, natural gas is decomposed in special conversion units before injection. Injection of converted gas is now being introduced into blast furnace practice.

Fuel oil and pulverized coal are also used as reducers. Fuel oil is a by-product of oil processing. It is a mixture of hydrocarbons which, like the natural gas, decomposes in furnace producing carbon monoxide and hydrogen. Fuel oil is relatively cheap, which renders its use advantageous.

Pulverized coal is not suitable for many industrial applications. Neither can it be charged into blast furnaces together with the burden since this would hinder furnace operation. However, it may be injected into the combustion zone in the lower part of a blast furnace. No deterioration in furnace performance is experienced and a saving on coke is effected.

CHAPTER 7

Blast Furnace Structure and Operation

7.1. Blast Furnace Structure

A blast furnace is a shaft lined on the inside with refractory materials. The furnace is charged from the top with burden materials, including coke, sinter, ore, fluxes, various additives, and supplied from the bottom with heated air and natural gas or other fuel which partly replaces

4*

the coke. The combustion of coke and natural gas in the blast furnace is incomplete due to an intentional shortage of blast, the products being carbon monoxide CO, hydrogen H_2, and nitrogen N_2. While rising, these gases heat the descending burden materials and reduce iron, manganese, and other elements from their oxides. Thus, a blast furnace operates on the *countercurrent principle*, the burden moving downwards and the gases upwards.

Profile and dimensions of a blast furnace. The inside configuration of a blast furnace is called its *profile*, or *furnace lines* (Fig. 19). Top *1*, or cylindrical part, ad-

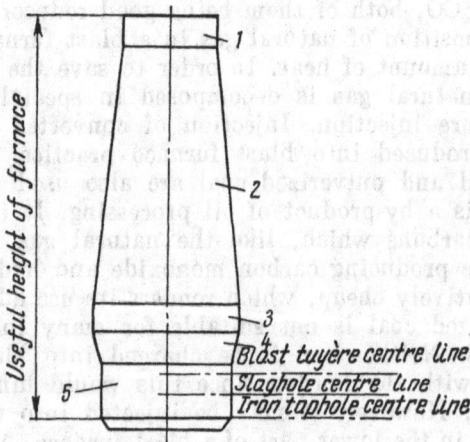

Fig. 19. Blast furnace lines
1—top, *2*—stack, *3*—body, *4*—bosh; *5*—hearth

joins stack *2*, which is the largest single part of the furnace. The stack gradually expands and passes into cylindrical body *3*, which is the widest part of the furnace. The lower part of the furnace—hearth *5*—has a smaller diameter and is connected to the body by bosh *4*. *Iron tapholes* (one or two) are located near the bottom of the hearth. The vertical distance between the iron taphole and the furnace top is the *useful height* of a blast furnace. Each element of the profile is sized with due regard for the particularities of operation of the blast furnace, the quality of raw materials and the fuel used.

An important characteristic of each blast furnace is its *useful volume*. The world's largest furnaces of a useful volume equal to 2,700-3,000 m³ are now being built in the USSR. The useful height of these furnaces exceeds 30 m, and the overall height is more than 80 m. The diameter of the hearth may be as large as 12 m, and that of the body may exceed 13 m.

Refractory lining, cooling arrangements and casing of a blast furnace. Blast furnaces are lined on the inside with refractory brick. Very high temperatures are developed in blast furnaces: from 300°C at the top to 2,100°C in the combustion zone inside the hearth. A large quantity of molten iron and slag is formed in the lower part of the furnace (the body, bosh, and hearth). The blast-furnace slag contains oxides which may attack the furnace lining. Therefore, the lining is made from materials which are stable to the high temperatures, mechanical loads, and corroding action of the slags. Such materials are *fireclay* and *carbon refractories*. Fireclay brick contains 50-60 per cent silica, 42 per cent alumina and up to 3 per cent ferric oxide. The melting point of fireclay is above 1,700°C. Brick of this composition resists the attack of slag as it reacts neither with acid nor with basic oxides. The total interior of a blast furnace is lined with fireclay brick of different classes (see Fig. 21). Lining is up to 1,000 mm thick at the top and up to 1,500 mm at the bottom.

Carbon blocks have recently come to be used for blast furnace brickwork. Slag and iron produce no corroding effect on the blocks as these are 90 per cent pure carbon. However, they are readily destroyed by water vapours and oxygen. In consequence, carbon blocks are solely employed to line the lower part of hearth, where they are protected from oxidation by a blanket of iron and slag.

Gas pressure in the hearth exceeds 0.3 MN/m², therefore blast furnaces are enclosed in steel jackets (*casings*), 24 to 40 mm thick.

The refractory lining is specially cooled to prevent its overheating and mechanical failure. To this end *water coolers* are inserted into the metallic casing and the refra-

ctory lining. These coolers are iron castings with cast-in
steel tubes, cooled by running water. If necessary, the
furnace casing can be sprayed with water on the outside.
Blast tuyères, slag tap holes, and hot-blast stove valve fittings are also water-cooled. A big blast furnace consumes
approximately 100,000 m³ of water daily. Evaporative
cooling is now being used on an ever increasing scale.

A blast furnace is erected on a massive foundation of
refractory concrete. Resting upon the foundation are the
main furnace pillars which support the bearing ring, or
the *mantle*. The mantle supports the columns which carry
the top arrangement. The mantle itself supports the brickwork and the steel structures of the upper part of the furnace.

7.2. Principle of Blast Furnace Operation

Burden materials are stored in bins *1* (Fig. 20) whence
they are conveyed to weighing hoppers and then to special
cars *4* termed *skips*. Skips loaded with burden are hoisted
by an electrical winch along inclined bridge *5* to the furnace top and dumped into the receiving hopper *6* which
prevents the escape of furnace gases into the atmosphere
during charging (Fig. 21).

The blast is supplied to the furnace from a blower via
a special blast main. Air is preheated to 1,000-1,300°C in
hot-blast stoves *7* before it is blown into the furnace. From
the stoves, the hot blast is admitted to the bustle pipe *8*,
and then injected into the furnace hearth via tuyères *9*.

As the fuel is burned in the hearth, the reducing gases
heat and reduce the burden materials. The gases emerge
from the burden at the top of the furnace and run off through
downcomers *10*. At the foot of the furnace the gas enters
cleaners *11* where it leaves the dust entrained from the furnace. The gas leaving the furnace contains up to 30% CO
and 6-7% H_2; 1 m³ of such gas when burnt in the air liberates 3,770 kJ of heat. Therefore, the purified blast furnace
gas is used as fuel.

Gravity forces the burden down, while gases ascend from
bottom to top. This causes the charge to be gradually heated and reduced, the resulting iron and slag trickling down

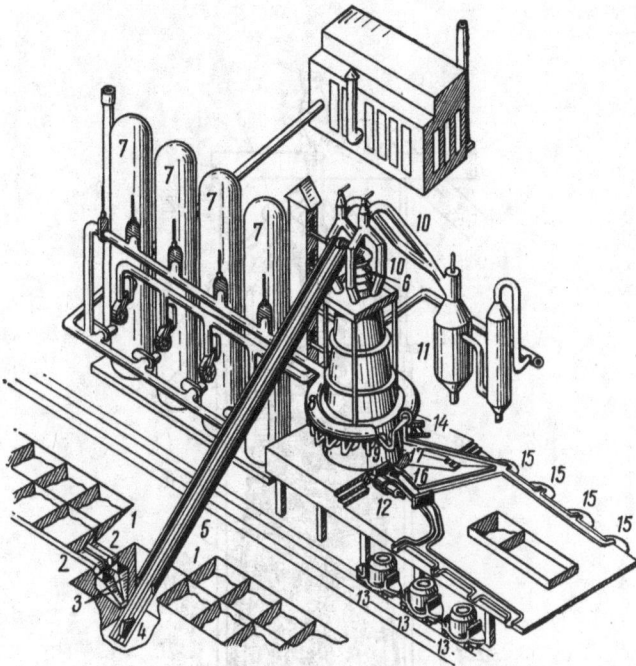

Fig. 20. Blast furnace plant

1 — bins; *2* — conveyors; *3* — weighing hoppers; *4* — skip; *5* — skip bridge;
6 — receiving hopper; *7* — hot-blast stoves; *8* — bustle pipe; *9* — tuyères; *10* —
gas uptakes and downcomer; *11* — gas cleaning devices; *12* — mud gun; *13* —
hot-metal ladle cars; *14* — stopper; *15* — slag cars; *16* — iron tapholes; *17* —
slag tapholes

into the hearth. The iron and slag are relieved from the furnace as soon as they accumulate, for which purpose an iron notch and a slag notch are provided. Iron notches are located level with the hearth bottom. After iron has been tapped, the hole of the iron notch is plugged with a refractory mass by means of a piston-type electric *mud gun*.

The cylinder of an electric mud gun (Fig. 22) is filled with a refractory mass. The cylinder terminates in a tip. When an iron notch is to be plugged, the tip of the mud gun is brought against the hole and the clay forced out of the cylinder plugs the notch.

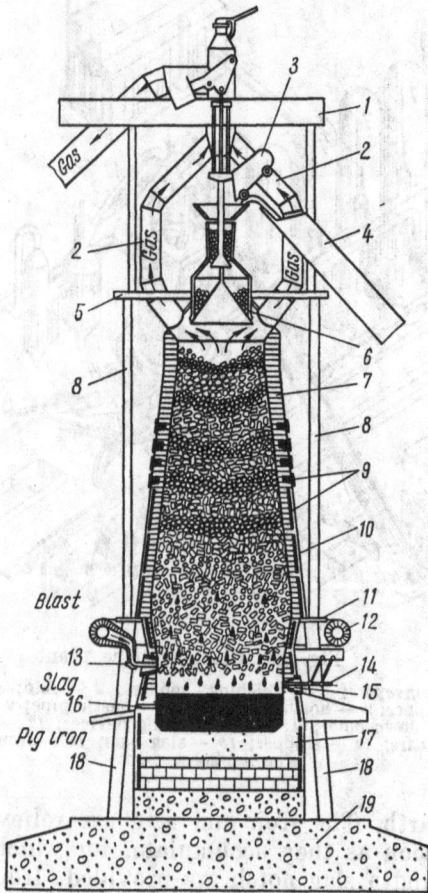

Fig. 21. Cross section of blast furnace

1 — framework; 2 — gas uptake; 3 — skip; 4 — skip bridge; 5 — top platform
6 — receiving hopper; 7 — fireclay brickwork; 8 — column; 9 — coolers; 10 —
casing; 11 — mantle ring; 12 — bustle pipe; 13 — tuyère; 14 — slag stopper
15 — slag taphole; 16 — iron taphole; 17 — carbon lining; 18 — main column
19 — foundation

To tap the iron, the iron notch is drilled, and the iron flows from the furnace along special launders into *hot-metal ladle cars*. A hot-metal ladle car is essentially a ladle lined on the inside with refractory brick and mounted on railroad bogies. Ladles are built in capacities up to 140 tons of iron.

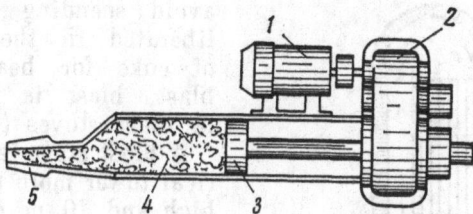

Fig. 22. Electric mud gun

1 — electric motor; *2* — gearing; *3* — piston; *4* — refractory mass; *5* — tip

A part of blast furnace iron is transported to a steel-making plant, and the remainder is cast into *pigs* in casting machines.

Slag notches are located between the level of the iron notches and that of the blast tuyères. The water-cooled nozzle of a slag notch is plugged by special stopper *14* (Fig. 20). Slag is conveyed to slag cars and transported to a slag granulation installation. Here molten slag is poured into water tanks and solidifies as round grains called granules which are then used in the production of construction materials. An alternative arrangement is to dump slag at cinder yards.

7.3. Auxiliary Arrangements

Blowers. The amount of air blown through a blast furnace reaches 8 million m³ per 24 hrs. The blast is supplied into the furnace under a high pressure (over 0.4 MN/m²).

The blast is supplied from a blowing machine, or *blower*. Atmospheric air travels along the blower shaft, the blades of the impeller revolving at a high speed; the blades throw the air towards the blower walls and thereby compress it. The compressed air then passes onto the blades of the next impeller and is compressed still further. This process is repeated several times, so that the pressure at the blower outlet is 4-5 times higher than atmospheric pressure. The blast is conveyed to the furnace via blast mains.

Blast stoves. In order to avoid spending the heat liberated in the burning of coke for heating the blast, blast is preheated in blast stoves (Fig. 23).

A blast stove is a cylindrical tower more than 40 m high and 10 m in diameter, lined with refractory brick. The tower houses a brick checker which offers thousands of through channels for the passage of gases. The channels are separated from one another by thin brick walls.

As the air supply pressure is in excess of 0.4 MN/m^2, the blast stoves are enclosed in steel casings.

The checker is heated by injecting atmospheric air and clean blast-furnace gas (sometimes with an admixture of natural gas) into the stove through a burner. The gas burns in

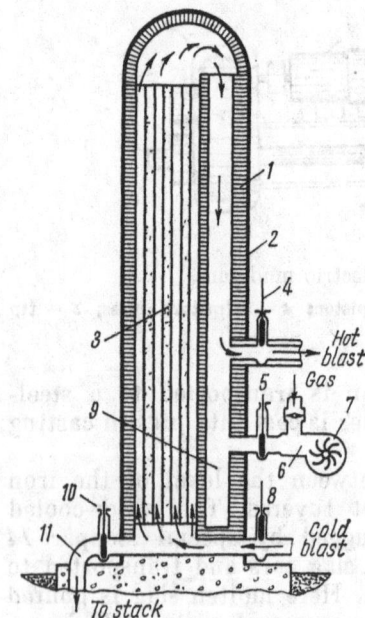

Fig. 23. Blast stove

1 — refractory brick; *2* — casing; *3* — checker; *4,5* — gate valves; *6* — burner; *7* — blower; *8* — gate valve; *9* — combustion chamber; *10* — flue valve; *11* — flue

a combustion chamber provided in the stove and the products of combustion rise to the dome, from where they are directed downwards via the checker channels, thus heating the checker brick to 1,300-1,400 °C. After passing through the checker, the gases are exhausted via a flue to a stack.

The checker is heated for 2-3 hours. Then the stove is cut off from the flue and the burner by valve *10* and slide gate *5*, respectively. Slide gates *8* and *4* are then opened to connect the stove to the blast furnace and the blower. The cold air from the blower enters the stove at the bottom and, while rising through it, heats up to 1,200-1,300°C. Under the dome, the air makes a turn and passes through the combustion chamber to the hot-blast line which conveys it to the bustle pipe of the furnace. From here, the blast is injected into the furnace hearth via tuyères 160 to 200 mm in internal diameter.

A modern high-capacity blast furnace consumes up to 5,500 m³ of blast per minute, and therefore the hot-blast stove cools as rapidly as in 1.0-1.5 hours. After this, the checker requires re-heating, and in the meantime the blast is heated in another blast stove.

A blast furnace is provided with four stoves: three of these are heated while the blast is supplied to the furnace via the fourth one.

Hot-blast stoves serve not only to heat the air. During a furnace shut-down the gases are sucked away from the blast furnace and burned in the stoves.

Gas cleaning. *Blast-furnace gas* contains up to 30% CO and up to 6% H_2 and has a relatively high calorific power. This gas is used as fuel for the blast stoves and open-hearth furnaces. However, the gas carries much dust, called *flue dust*. Gases with so high a dust content cannot be used in heating units, as the latter may be chocked up.

The gas is cleaned in devices, adjoining the blast furnaces (see Fig. 20). It first enters the primary dust catcher of a diameter in excess of 10 m, where its speed slows down rapidly. The heavier dust particles continue to move downwards by inertia and settle down on the dust catcher bottom.

From the primary dust catcher the gas is conveyed via a flue to the bottom of a *scrubber*, accommodating water-sprayed wooden grids. The gas is cleaned from the remaining dust as it rises along the wet grids. A clean blast-furnace gas is then discharged to the clean-gas main and conveyed to consumers.

CHAPTER 8

Blast Furnace Process

Physical, chemical and aerodynamic processes take place in a blast furnace. The course of each of these is interrelated with that of the others.

As the charge descends towards the hearth, it undergoes a number of consecutive changes and transformations. We shall examine the course of the principal processes in the order of their occurrence over the height of a blast furnace. Different processes occur at different levels of the furnace, this being due to the fact that the temperature, composition, and pressure of gases change as they ascend from zone to zone. The nearer to the hearth, the higher the temperature and pressure of gases and the greater the content of carbon monoxide.

The principal processes taking place in a blast furnace are as follows:

(1) evaporation of moisture;

(2) evolution of volatiles;

(3) decomposition of carbonates;

(4) reduction of oxides into metallic iron and other elements;

(5) carburization of iron and formation of pig iron;

(6) slag formation;

(7) combustion of fuel.

8.1. Decomposition of Charge Materials

Evaporation of moisture. The temperature of gas at the furnace top, where it evolves from the burden, is in excess of 200 °C, being still higher inside the bed of materials. At these temperatures evaporation of moisture from the charge starts as soon as the materials fall into the furnace. Evaporation of free, or hygroscopic, moisture is very rapid. However, the burden may also contain hydrate (i.e., chemically bound) moisture. Brown ores are high in hydrate moisture. Its removal requires a high temperature (up to 800 °C) and long time.

Evolution of volatile matter. Not all volatile substances (hydrogen, nitrogen, and others) are removed from the coke in the process of coking. This process goes on in the blast furnace, coming to an end in the high-temperature zones.

Decomposition of carbonates. Carbonates introduced into blast furnaces with the fluxes are limestone $CaCO_3$ and magnesite $MgCO_3$. Sometimes iron-carbonate ore $FeCO_3$ or manganese carbonate ore $MnCO_3$ is added to the burden. Iron carbonate has the lowest decomposition temperature of these salts (520 °C), those of $MnCO_3$ and $MgCO_3$ being somewhat higher. The highest decomposition point is that of limestone (900-920 °C). Limestone decomposes according to the following reaction:

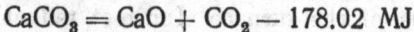

$$CaCO_3 = CaO + CO_2 - 178.02 \ MJ$$

The negative value of the quantity of heat denotes that heat is absorbed in the reaction.

Decomposition of other carbonates follows a similar pattern: salts dissociate into metallic oxides and carbon dioxide CO_2. All these reactions absorb heat, therefore it is advisable to avoid introducing carbonates into the burden.

8.2. Reduction of Iron

Iron is reduced in the blast furnace by successive separation of oxygen from the oxides, i.e., by formation of new oxides with a lower oxygen content: first Fe_2O_3 is reduced to Fe_3O_4, then to FeO and, finally, to pure iron.

Oxygen of the oxides is taken away by the reducers, namely, carbon monoxide, hydrogen, and carbon. The principal reducer is carbon monoxide. The reduction of iron from Fe_2O_3 consists of the succession of reactions below:

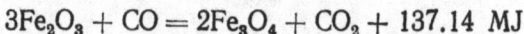

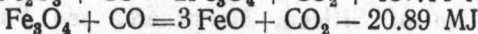

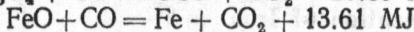

$$3Fe_2O_3 + CO = 2Fe_3O_4 + CO_2 + 137.14 \ MJ$$
$$Fe_3O_4 + CO = 3 FeO + CO_2 - 20.89 \ MJ$$
$$FeO + CO = Fe + CO_2 + 13.61 \ MJ$$

Carbon dioxide evolves in all the three reactions. This is called *indirect reduction*. It is possible in the upper zones of the furnace, at temperatures below 1,000 °C. Carbon dioxide cannot exist in the presence of carbon at higher

temperatures, as it reacts with the carbon of the coke re-
producing carbon monoxide:

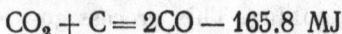

$$CO_2 + C = 2CO - 165.8 \ MJ$$

This reaction absorbs a great amount of heat, and the
solid carbon of the coke is consumed in the process.

In the lower part of the furnace, coke lumps are immersed in molten slag which causes an intimate contact of the
coke carbon with the slag oxides. In this case, the reducer
is coke carbon:

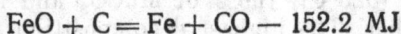

$$FeO + C = Fe + CO - 152.2 \ MJ$$

The reaction in which carbon monoxide is one of the re-
sultants is termed *direct reduction*. It proceeds mostly in
the lower part of the furnace. Indirect reduction liberates
heat, while direct reduction absorbs it. Therefore, the furnace should be operated in a manner minimizing the progress of the direct reduction. Under normal conditions, the
direct reactions account for approximately 30-50 per cent
of the reduction process.

Before natural gas has come to be used, the principal
reducer in the furnace was carbon monoxide. Its content
in the hearth gas reaches 40 per cent, while that of hydrogen does not exceed 1-2 per cent. When natural gas is injected, the hydrogen content in the hearth gas goes up by
several times, thus the reduction of iron oxides by hydrogen grows in importance. The process gives rise to the formation of water vapour:

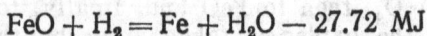

$$FeO + H_2 = Fe + H_2O - 27.72 \ MJ$$

All the reducing reactions proceed the faster, the greater
the content of reducer gases (carbon monoxide, or hydrogen) and the higher the temperature.

8.3. Reduction of Silicon, Manganese, and Phosphorus. Desulfurization of Iron

Reduction of silicon. Silicon is reduced in blast furnaces from silica contained in the slag by the solid carbon
of the coke:

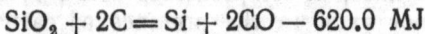

$$SiO_2 + 2C = Si + 2CO - 620.0 \ MJ$$

The reaction proceeds at very high temperatures (above 1,050 °C) with the absorption of heat. Such temperatures are only possible in the lower part of the furnace. The greater the silicon content in the iron, the higher the temperature in the furnace and the greater the consumption of coke per ton of iron produced. High silica in the slag (acid slag) is beneficial as regards the reduction of silicon.

Reduction of manganese. The properties of manganese resemble those of iron. Its reduction proceeds successively, from oxides rich in oxygen to those leaner in it:

$$MnO_2 \rightarrow Mn_2O_3 \rightarrow Mn_3O_4 \rightarrow MnO \rightarrow Mn$$

Reduction to manganous oxide MnO is rapid and ends in the stack. Manganous oxide is difficult to reduce, and this reaction goes on at lower furnace levels, where the process is facilitated by higher temperatures. Here it is reduced directly by the solid carbon of the coke:

$$MnO + C = Mn + CO - 258.9 \text{ MJ}$$

Thus, the reduction of manganese, similarly to that of silicon, requires much coke. However, in contrast to silicon, the reduction of manganese presupposes a slag low in silica (basic slag). About one half of the manganese of the burden appears in the iron, the rest goes out with slag.

Reduction of phosphorus. All of the phosphorus fed to a blast furnace, is reduced and penetrates into iron. A high temperature is necessary to reduce phosphorus. In normal operation, phosphorus does not go into slag, which renders its removal from iron impossible.

Desulfurization of iron. Sulfur gets into the furnace with the ore, sinter, and fluxes, but its chief vehicle is coke. Coke sulfur is oxidized in the hearth by the oxygen of air to give the sulfur dioxide SO_2. As this gas ascends in the furnace stack, it comes into contact with incandescent lumps of coke and is reduced to free sulfur:

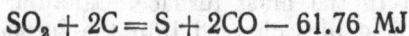

$$SO_2 + 2C = S + 2CO - 61.76 \text{ MJ}$$

Free sulfur reacts with iron to give ferrous sulfide FeS which readily dissolves in the iron and resists passage into slag. Ferrous sulfide in solid iron and steel has a very adverse effect on their properties. Sulfur may be eliminated from iron by transferring it to a compound insoluble in iron. Such a compound is calcium sulfide CaS, which may be obtained by the following interaction between CaO of the slag and FeS of the iron:

$$CaO + FeS = CaS + FeO$$

Ferrous oxide is then reduced by solid carbon:

$$FeO + C = Fe + CO$$

Both reactions absorb great amounts of heat and, therefore, require much coke. Sulfur is eliminated from iron the better, the higher is the calcium oxide content in the slag (basic slag).

8.4. Carburization of Iron and Formation of Pig Iron

Pure iron melts at 1,527°C. However, if carbon is dissolved in it, the melting point of the resulting alloy is substantially lower.

Depending on the content of other elements, a blast furnace iron may contain from 3 to 7.5% C. The commonest variety of iron is the steelmaking pig iron, used for the manufacture of steel. It assays 3.7-4.9 per cent carbon and its melting point is about 1,150°C.

Iron carburization in a blast furnace is due to carbon monoxide:

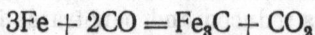

$$3Fe + 2CO = Fe_3C + CO_2$$

The resulting iron carbide Fe_3C dissolves in iron and sharply decreases its melting point. The resulting pig iron trickles down into the hearth.

8.5. Formation of Slag

Slag is formed by the oxides of silicon, calcium, magnesium, aluminium, manganese, and iron, which have failed to be reduced in the blast furnace. Calcium, magnesium, and aluminium oxides charged into the furnace pass entirely into slag. These oxides are hard to reduce and usually are not reduced in the process of blast-furnace smelting. Silica almost entirely enters slag. Only a small portion of silicon is reduced from slag and appears in the iron.

About a half of the manganese is reduced and passes to iron, while the unreduced oxides remain in the slag. Almost all of the iron is reduced and enters pig iron, so the slag is low in ferrous oxide.

The oxides of silicon, calcium, magnesium, and aluminium which make up the bulk of the slag have very high melting points. However, if they are mixed in certain proportions, the mixture may melt at lower temperatures. This is what actually occurs in blast furnaces: as the charge descends to the body or to the lower portion of the stack, such mixtures fuse together and then melt to give slags, termed *primary slags*. These are rich in ferrous oxide, which makes the slag very fluid.

As the primary slag trickles down, it heats up and gradually dissolves the remaining unreduced oxides with the effect that the content of CaO and MgO goes up. The percentages of FeO and MnO in the slags decrease as they are reduced by the carbon of the incandescent coke and go over to iron.

Coke ash composed mainly of silica passes into slag at the tuyère level. Besides, pig iron components, such as iron, manganese, silicon, are partly oxidized by the oxygen of the blast in the combustion zone. The resulting oxides also pass into the slag. However, as they trickle down into the hearth, they come into contact with incandescent coke and are almost entirely reduced again.

In the hearth, the slag floats over the iron since it is the lighter of the two. As iron droplets pass through the slag blanket, they lose their sulfur (see Para 8.3). Now the slag is completely formed. In contrast to the primary slag, this is termed the *final slag*.

The processes of slag formation are most important for blast furnace operation. The blast furnace throughput and the quality of iron depend on an adequate slag practice. Slag composition is selected to fit the kind of iron to be smelted as this composition determines the slag fluidity and ensures an adequate and timely removal of sulfur from the iron.

Slags obtained in the production of steelmaking iron usually contain 48% CaO, 40% SiO_2, 5% MgO, 5% Al_2O_3, and small amounts of ferrous and manganous oxides. When foundry iron is smelted, the slags are somewhat leaner in calcium oxide, but richer in silica.

Timely tapping of slag from the furnace is of major importance. A delay may cause the slag to rise to the tuyère level and block the blast outlets, thus slowing down the rate of smelting, reducing the furnace throughput. and even causing an accident (in case a tuyère burns through).

8.6. Combustion of Fuel

Coke lumps are the only solid material reaching the tuyère level, where all the other materials are liquid.

Air is injected into the hearth through blast tuyères which may number from 14 to 20 depending on the size of the furnace. The blast bursts into the hearth at a velocity of 100-150 m/s, so that the coke before the tuyères spins in a whirl, this providing an intimate contact between the coke and the blast. Coke carbon burns according to the reaction

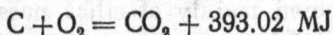

$$C + O_2 = CO_2 + 393.02 \text{ MJ}$$

i.e., carbon dioxide is formed and a great amount of heat liberated. However, at a distance of a mere 1 m from the tuyère there is no more oxygen in the gas, as all of the oxygen has been consumed. In consequence, the incandescent coke reacts with the carbon dioxide as

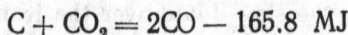

$$C + CO_2 = 2CO - 165.8 \text{ MJ}$$

This reaction produces carbon monoxide and consumes a considerable amount of the heat generated by coke conversion into CO_2.

Thus, the gas above the tuyères is composed of carbon monoxide and nitrogen. The latter is introduced with the blast and does not participate in any reactions in the furnace.

Burning of coke to CO_2 liberates more heat than is consumed in the reaction between CO_2 and the coke. The balance is sufficient to maintain a temperature of 1,700-2,000 °C in the combustion zone. The heat of fuel combustion goes into heating both the iron and the slag and reducing the burden. The ascending high-temperature gases heat the descending materials and reduce oxides contained therein.

As combustion turns solid coke into gases and only a small quantity of ash remains which enters the slag, a void is formed in the combustion zone.

Formation of the void in the lower part of the furnace is also due to the melting of materials. This void is filled by the descending burden. Since the combustion of coke and melting of materials are continuous, the materials also move downwards at a uniform rate. Blast furnace smelting proceeds the faster, the greater the rate of descent of the burden, i.e., the larger the space vacated by the combustion of coke. The quantity of coke burned is the greater, the larger is the amount of blast blown into the furnace per unit time.

In recent years, coke has been partially replaced in many blast furnaces by the natural gas composed chiefly of methane CH_4. Natural gas is injected into the hearth through the blast tuyères. As blast and methane mix in the combustion zone, the methane decomposes and partly oxidizes, giving rise to carbon monoxide CO and hydrogen H_2. The latter is a very strong reducer. As it rises together with carbon monoxide, it reduces most of the burden oxides. Therefore, a smaller quantity of carbon monoxide is required from the coke, and thus a smaller amount of coke itself is necessary.

Decomposition of the natural gas before the tuyères calls for a supply of heat, and this is why the temperature in the combustion zone is somewhat below the usual values, the same as on injection of fuel oil.

CHAPTER 9

Blast Furnace Smelting Products

The products of blast furnace smelting are iron, slag, and gas. Blast furnaces produce irons for various applications.

Iron. Steelmaking iron, or *pig iron*, is intended for conversion to steel. This grade of iron is brittle, takes the mould impression poorly, and is unsuitable for the manufacture of as-cast machine parts. It contains 0.3-1.2% Si

5*

and up to 2.5% Mn. Sulfur content should be kept as low as possible, since its removal during processing to steel is very costly. Usually it assays to less than 0.060 per cent sulfur. Upon tapping, pig iron is directly transported to steelmaking plants.

Cast *iron* is intended for casting various parts; it is very hard and sufficiently strong. Its sulfur content should be lower than that of pig iron owing to the presence of sulfur which impairs the strength of finished parts. Cast iron has good casting properties owing to its high silicon content (1.3-4.7 per cent). Manganese is usually below 1.0 per cent.

Along with the cast iron, *malleable* and *roll irons* are used for the manufacture of critical parts. A feature of these irons is their high strength; in contrast to the common-grade irons, they are not brittle.

Special irons, called *ferroalloys*, such as *ferrosilicon* (containing 9-15% Si) or *ferromanganese* (containing 70-80% Mn), are also smelted in blast furnaces.

Iron to be supplied solid is run from the blast furnace to a pig machine where it is cast into moulds. Solidified ingots, termed *pigs*, are loaded into railway cars and shipped to consumers.

Slag from blast furnaces is flushed into slag cars, i.e., special steel pots mounted on railway carriages. Filled slag cars are conveyed to water tanks, and slag is poured into the latter. As the slag falls into water, it disintegrates into round gravel-like droplets and solidifies.

As the slag contains a large amount of lime, silica, and alumina, it may subsequently be used for the manufacture of cement or other construction materials.

As has already been mentioned, blast furnace gas contains up to 30 per cent carbon monoxide and 4-7 per cent hydrogen. The combustion of 1 m^3 of such a gas liberates up to 3,770 kJ of heat so that upon dust removal it may be employed as fuel. It is used to heat hot-blast stoves, open-hearth furnaces and rolling plant heating furnaces. Several million cubic metres of gas are exhausted daily by blast furnaces, thus its use as a source of heat effects a sizable saving on fuel.

CHAPTER 10

Blast Furnace Practice

10.1. Blowing In and Out

A new blast furnace is lined on the inside with refractory brick, bonded by liquid fireclay mortar. Before iron smelting begins, the new lining should be dried, otherwise it will crack in the process of rapid heating. The lining is dried at a slow rate, over a few days or even weeks, by burning wood or gas to gradually raise the temperature inside the furnace.

Once the furnace is dry, it is charged with the *blow-in charge*. Prior to this, the lining is cooled down to 50°C in order to prevent a premature ignition of the coke in the furnace. The blow-in charge is so proportioned as not only to obtain an iron of a specified composition, but also to heat the furnace. This is why the blow-in charge is always high in coke.

The burden is dropped into the blast furnace from the top and the first portions of materials fall from a great height. In order to prevent damage to the brickwork of the hearth and the bosh, the lower part of the furnace is filled with wood. Charged atop the wood are then coke, limestone, and sinter.

When the furnace is filled to capacity with the burden materials, it is *blown in* by admitting hot blast from the blast stoves. The coke ignites and the resulting gases ascend to gradually heat and then reduce the oxides in the burden materials.

Thus starts the life of a new blast furnace. After 7 to 10 years of operation, the blast furnace wears out and "ages": furnace casing gets out of order, the equipment grows obsolete and calls for replacement. The furnace is shut down for overhaul, the stoppage involving its *blowing out* to remove the materials it holds. To this end, the furnace is supplied with the blast, but its charging is discontinued. The coke in the furnace gradually burns out, the materials melt and go down to the tuyère level. The temperature in the furnace rises sharply, since the gases are not cooled by the freshly charged materials. To prevent overheating of or damage to the upper structures of the furnace, water is fed inside the furnace to cool the gases.

As the burden approaches the tuyère level, iron and slag are tapped from the furnace, the blast supply is cut off, water is poured instead of the blast through the tuyères to cool the incandescent coke and the molten iron and slag remaining in the furnace. After this an opening is cut in the casing and the rest of the materials are removed from the furnace.

The interval between the blow-in at the start of furnace operation and the blow-out prior to an overhaul is termed the campaign or *running time* of the furnace. The better a furnace runs and the more efficiently the furnacemen operate it, the longer is the campaign of a furnace. There are blast furnaces that operate without overhauls for more than 10 years.

10.2. Normal Operating Conditions of Blast Furnaces

The chief aim of blast-furnace smelting is to produce iron of a required composition at minimum cost and least wear of the furnace equipment. The basic parameters:

(1) burden composition and charging schedules

(2) quantity, pressure, and temperature of the blast, and

(3) tapping schedules

are chosen to fit the grade of the iron smelted.

The composition and properties of burden materials are not invariable, but change, this affecting the smelting conditions. The task of furnacemen is to timely notice these changes and adjust the operating parameters accordingly.

The experience of foremost plants shows that a high performance is obtained when the rate of the burden descent is uniform. This ensures the least consumption of coke, since in a furnace operating at a uniform rate the gases permeate the charge evenly and their reducing energy is used to the maximum. On the other hand, the lower the coke consumption, the larger quantity of iron may be smelted on the same amount of blast. The simplest way to ensure a steady furnace operation is to supply a small quantity of blast. But the output of iron will then be adequately small. The art of the furnacemen is to supply the maximum quantity of blast compatible with a uniform rate of burden descent. If the limit of blast is exceeded, the powerful gas streams may blow channels in the charge, through which the bulk of the gases pass. Their energy is then poorly used, since gas distribution across the column of burden materials is erratic; the consumption of coke goes up, and the output of iron diminishes despite the high amount of blast.

The steadiness of furnace operation may be judged by the even descent of the burden, the character of variations in pressure and amount of blast, the glow of coke in front of tuyères, and by other symptoms.

When controlling the rate of the burden descent, account should be taken of the thermal conditions inside the furnace. For example, when the burden slips jerkily, the heat balance is bound to be affected adversely. Therefore, proper measures should be taken simultaneously with the slip correction to ensure an adequate heat input to the furnace.

This may be achieved by raising the consumption of coke and/or the blast temperature, by decreasing the moisture content in the blast, etc. The choice of the method depends on specific conditions, as the furnace responds differently to different corrective actions. For example, uneven burden descent should never be remedied by raising the blast temperature. It is preferable to adjust the charging schedule so as to facilitate the passage of the gases and to increase, at the same time, the consumption of coke. Variations in heating conditions, in turn, affect the burden descent.

If the heat input is increased sharply, the furnace runs harder. Then the blast temperature should be temporarily reduced in order to lower the heat input and thus to improve the rate of burden descent.

It follows then that the blast furnace operation is an intricate matter, requiring much knowledge and experience. It is necessary to analyze the changes in furnace behaviour, to anticipate the effect of these changes upon furnace operation and to make the adjustments accordingly. Blast furnacemen are aided by numerous instruments and automatic devices in solving these complicated problems.

10.3. Automation of Blast Furnace Operation

Variations in most of blast furnace parameters are continuously monitored by instruments and recorded on charts. The readings of the instruments serve as the basis for evaluating the furnace operation. Many of the instruments, besides ensuring data readout, also trigger off automatic circuits which automatically maintain the operating parameters at preset levels. These circuits control blast temperature and moisture content, pressure of gases at the furnace top, temperature in blast stoves, etc.

The highest level of automation in blast furnaces has been achieved in the most difficult job, that of burden charging in the furnace. The participation of an operator boils down to the specification of the burden composition and the sequence of charging. The rest of the work is performed by automatic circuits which control the operation of hundreds of mechanisms.

The goal that confronts the blast furnace practice at present is not only to automatically maintain the smelting parameters at the prescribed levels, but also to control the smelting fully automatically, without the participation of man. However, the blast furnace processes have yet been studied insufficiently, and, therefore, their automatic control is still a certain distance away.

CHAPTER 11

Means for Increasing Blast Furnace Productivity

The productivity of a blast furnace is a function of the quantity of the blast supplied to the furnace and the relative consumption of coke, which, in turn, both depend on careful preparation of the raw materials.

The greater the amount of small particles and dust in the burden, the smaller the number of gas channels in the bed of burden and the lower amount of blast can be blown into the furnace. It was shown in Chapter 8 that blast intensification brings about a higher furnace productivity. Therefore, the burden should consist of sufficiently large lumps and be free from dust.

Of major importance is the grade of the burden. The higher the iron and the lower the gangue content, the smaller the relative coke consumption, i.e., the quantity of coke necessary to smelt 1 ton of iron. It is evident that a lower relative consumption of coke is paralleled by a higher furnace productivity.

Suppose that two similar blast furnaces operate side by side and consume equal amounts of blast. This means that the furnaces burn the same quantity of coke per unit time. However, in one of the furnaces the relative coke consumption is 0.5 ton per 1 ton of iron, and in the other it is 0.6 ton per 1 ton of iron. Hence, when both furnaces have burned 500 tons of coke each, the first furnace would have smelted 1,000 tons of iron, and the second one, a mere 833 tons. Therefore, the lower the relative consumption of coke, the higher the productivity of the blast furnace. In consequence, all appropriate means should be employed to bring down the coke consumption.

Examine in greater detail the ways towards increasing furnace productivity.

11.1. Preparation of Raw Materials

Size of burden materials. We know already that a dusty burden impairs the operation of a blast furnace, and this is why fines should be removed from the burden before the latter is charged into the furnace. But dust may also result from disintegration and abrasion of burden lumps in the furnace. This requires that the burden materials— coke, ore, sinter, pellets — are mechanically strong.

Grade of burden. Materials smelted in blast furnaces should be as rich in iron as possible. Gangue removal reduces the useless consumption of energy for heating and slagging the gangue and, thereby, brings down the relative consumption of coke. Moreover, elimination of the gangue improves the gas permeability of the stock since a smaller amount of slag is formed in the furnace; this, in turn, allows more blast to be supplied to the furnace.

Self-fluxing sinter. Decomposition of raw limestone consumes a certain quantity of heat the provision of which in a blast furnace necessitates a greater amount of coke than in a sintering machine, as in the latter all of the carbon burns to carbon dioxide to give the maximum heat effect. In a blast furnace less than half of fuel carbon is burned to carbon dioxide, so that its calorific power is used incompletely. It is in all respects preferable to introduce only sintered limestone into the burden. In this case blast furnace heat will not be consumed for decomposition of the limestone, coke consumption will decrease, and furnace productivity will rise.

11.2. Cutting Down the Specific Fuel Consumption

Heating of blast. The smelting of 1 ton of iron in a blast furnace requires some 2,000 kg of air. If this air is preheated, the furnace will be supplied additionally with a great amount of heat. Air is preheated in blast stoves whose checker is heated by burning blast furnace gas (see Chapter 7).

It is evident that it is more advantageous to heat the required air in blast stoves than to produce the necessary heat in the blast furnace proper. The heating of air to, for example, 1,000°C requires the same amount of heat both in the furnace and in the blast stove. But the quantity of fuel consumed is different: 100 units in the furnace versus 80 units in a hot-blast stove. The cause of this is the incomplete use of the calorific power of the fuel in a blast furnace that has been mentioned above.

In current practice, furnace blast is preheated to 1,000-1,200°C. This accounts for about 20 per cent of the total heat necessary for the smelting.

Maximum use of the energy of the gas stream. It is essential that an intimate contact be ensured in the blast furnace between the gases and the burden. The better the contact, the greater the amount of heat transmitted by the gases to the burden and the better the reduction of ore and sinter. However, it is unreasonable to expect that such a contact between the gases and the charge materials could be established of itself. The gases pass through the larger channels and flow more readily through the bed of coke than through the bed of ore or sinter. But as it is the ore and sinter that need reduction, means should be devised to cause the gases to flow through the accumulations of ore and sinter.

Besides, the movement of the gases and that of the ore materials should be so arranged as to effect greater reduction of oxides in the upper furnace zones, where the temperature is below 1,000 °C. As is known from Chapter 8, indirect reduction of iron is possible at these temperatures, with carbon monoxide oxidizing to carbon dioxide. The indirect reduction requires less heat.

If the material has failed to be reduced in the upper part of the furnace, it is reduced by direct reactions in the lower zones. The resulting carbon dioxide reacts with incandescent coke and is converted to carbon monoxide, the reaction absorbing large amounts of heat, i.e., necessitating additional coke.

It is essential to channel the gases and control the burden descent in a manner limiting the direct reduction.

11.3. Intensification of Blast Furnace Smelting

Blast enrichment in oxygen. A blast furnace may be supplied with only a limited amount of blast to oxidize the coke. The greater the amount of coke burned, the higher is the productivity of the furnace under equal conditions. How can a greater quantity of coke be burned with the same amount of blast? Atmospheric air contains 21 per cent oxygen and 79 per cent nitrogen. If oxygen content in the blast is increased, a greater amount of coke will be burned with the same quantity of blast.

Oxygen obtained in special installations is added to the blast. Oxygen content in enriched blast ranges from 26 to 30%. Operation with oxygenated blast, however, poses new problems in connection with a sharp increase in the temperature of the hearth to about 2,100 °C. At this temperature the slag oxides evaporize. Their vapours rise, cool down and plug the gas channels. The resistance of the bed of burden to the flow of gases increases, burden descent becomes erratic or stops altogether; this condition is known as "scaffolding".

The cause of the sharp increase in the hearth temperature is as follows. 1 m³ of atmospheric air (composed of 21 per cent oxygen and 79 per cent nitrogen) burns about 225 g of coke carbon. In the process, a certain quantity of heat is liberated and carried away by the combustion gases. If oxygen content is raised to 30 per cent, the quantity of burned coke and liberated heat will increase by almost 50 per cent, whereas the volume of the evolving gases will increase by a mere 6 per cent. In consequence, the gases will be heated to a higher temperature. When the blast is oxygenated, the temperature rise is prevented by injecting natural gas or steam into the furnace. The decomposition of these gases consumes heat and thus brings the temperature down.

At present, blast enrichment in oxygen is widely used in blast furnace practice. An increase in oxygen content by 1 per cent provides, on the average, a productivity raise of 2 per cent.

Humidification of the blast. One of the means for intensifying the blast furnace operation is to inject steam. In the furnace, the steam decomposes:

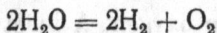

$$2H_2O = 2H_2 + O_2$$

Two volume units of steam give two units of hydrogen and one unit of oxygen. This oxygen burns an additional quantity of coke. Thus, the supply of steam is, in a sense, equivalent to blast enrichment in oxygen. Moreover, the nascent hydrogen accelerates the reduction processes. On the other hand, decomposition of steam consumes much heat, and this causes a relative increase in coke consumption. What is more, steam impairs slag fluidity and damages the carbon lining of the hearth. This is why only a limited quantity of steam may be injected into blast furnaces.

Increase of top pressure. Blowing of great amounts of blast into a blast furnace increases the lifting power of the gases and hampers burden descent. The lifting power is in direct relation to the quantity of the gases and the square of their velocity: as gas velocity doubles, the lifting power increases four times.

Gas velocity may be decreased by maintaining a higher gas pressure in the furnace. When the pressure of a moving gas is increased (for example, doubled), its velocity drops (is halved), provided the flow rate remains constant.

Gas pressure in the furnace is raised by obstructing the flow of gases. When special throttle valves downstream of the gas cleaners are closed, gas pressure upstream of the valves, including the pressure inside the furnace, goes up. The usual practice is to have the gauge pressure of gases at the furnace top equal to 0.11-0.18 MN/m^2, the pressure in the hearth reaching 0.23-0.30 MN/m^2. This raise in the pressure permits a greater amount of blast to be supplied and furnace productivity to be increased by 20-25 per cent.

11.4. Blast Furnace Performance

A most important performance characteristic of a blast furnace is its productivity. Modern blast furnaces of a useful volume of 2,700 m^3 and over smelt up to 6,000 tons of iron per day.

Smaller blast furnaces produce less iron, and there is no point in comparing their productivities to those of larger furnaces. However, their performances may be compared by using a different characteristic, the useful volume utilization (UVU) factor. This factor is the ratio of the useful furnace volume in m^3 to the amount of iron in tons smelted per 24 hours.

For example, the UVU factor of a 2,000-m^3 furnace which smelts 4,000 tons of iron per 24 hrs, is 0.5 m^3/ton. The greater the quantity of iron produced, the lower the UVU factor. Many furnaces in the USSR have a UVU factor below 0.5, and in some furnaces its value is as low as 0.4.

An equally important performance characteristic is the amount of coke (in kg) consumed in the smelting of 1 ton of iron. It is apparent that the lower the relative coke consumption, the cheaper is the iron. In modern practice, coke consumption in most of blast furnaces ranges from 400 to 600 kg per 1 ton of iron.

Blast furnaces of the Soviet Union are highly efficient
plants with a high level of automation and mechanization
and foremost production techniques. Also, Soviet blast
furnace plants show a high productivity of labour: up to
5,000 tons of iron per worker per year. The world's record
of daily output of iron per furnace belongs to one of the
blast furnaces at the Cherepovets Works.

CHAPTER 12

Direct Manufacture of Iron from Ores

In ancient times, iron was manufactured in bloomeries.
Their height did not exceed 1 or 2 m. The furnace was
charged with ore and charcoal, and the blast was supplied
from below by bellows. Because of the low height of the
furnace, the outgoing gases had a temperature in excess
of 1,000°C as they had no time to cool. Therefore, the direct
reduction predominated in bloomeries, accompanied with
the formation of carbon monoxide. As is known, this type
of reduction requires much fuel.

In modern blast furnaces, the reduction process takes
place at the higher levels, and the slag forms at the bottom. Therefore, few iron oxides reach the slag-formation
zone. In bloomeries, on the contrary, reduction and slag
formation proceeded simultaneously owing to the very high
temperatures in the furnace top.

Because of this, bloomery slags were rich in iron oxides
which hindered the reduction of silicon, manganese, and
phosphorus and the carburization of iron; thus, the product obtained was almost pure iron with a melting point
of about 1500°C. The temperature in the bloomery did
not reach this point and the iron did not melt, but formed
loose granular mass termed *sponge iron*. The furnace had
to be cooled down to recover the solid sponge iron. High
fuel consumption, frequent stoppages, and small sizes of
bloomeries conditioned their low output.

As the bloomeries grew in size, the temperature in their
upper zones diminished. The reduction zone separated from
the slag formation zone. The content of ferrous oxide in

the slag decreased, and the reduced silicon, manganese, phosphorus now passed from the slag into the metal. The metal became saturated with carbon, and it was no longer pure iron, but *pig iron*. This was the origin of the blast-furnace process. However, pig iron had a high carbon content, and was therefore brittle and only rarely suitable for the manufacture of various articles. Pig iron had to be converted to steel by removing carbon and other impurities. The necessary techniques were soon found, which gave birth to steelmaking. This marked a departure from a direct manufacture of iron and steel from ore and its substitution by a multi-stage process: ore → pig iron → steel.

However, methods of direct conversion of ores into iron are now again being developed.

Manufacture of iron in rotary furnaces. A mixture of ore and fuel (Fig. 24) is charged into a rotary furnace.

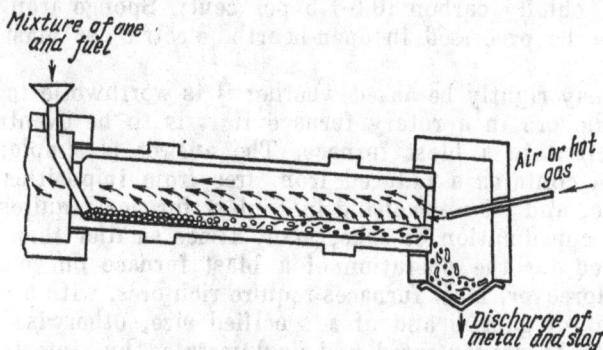

Fig. 24. Direct manufacture of iron in rotary furnace

Coarse ore is crushed into lumps 10 mm in size. Coke fines, anthracite or coal powder may be used as fuel. The furnace is slightly inclined and revolves slowly, so the ore and fuel slide gradually to the right and downwards. Air or combustible gas is supplied into the furnace from the right, and the ore charge gradually heats up. The higher the temperature, the greater the rate of ore reduction by the fuel carbon. As carbon monoxide is formed in the furnace on combustion of fuel and gas, it reduces some amount of iron oxides. Carbon monoxide, resulting from iron reduction by fixed carbon, burns by combining with the oxygen of the blast, and heats the furnace. The greater part of iron is reduced at 900-1,000°C. At this temperature the iron cannot melt, but it cakes into a *sponge*. The gangue softens at 1,000°C and melts at higher temperatures. This paste-like slag envelops the sponge iron and is discharged together with it at the end of the furnace.

The product is a semi-liquid slag with inclusions of sponge-iron lumps up to 200 mm in size. After cooling by water or air, this mass is charged into a crusher where the slag is ground to powder and separated from sponge iron by screening. However, the screened fine slag carries small particles of iron which are recovered in a magnetic separator (see Chapter 6).

The resulting sponge iron contains (after separation from slag) up to 98 per cent iron and a small amount of impurities, chiefly carbon (0.5-1.5 per cent). Sponge iron may further be processed in open-hearth, electric, or blast furnaces.

It may rightly be asked whether it is worthwhile to process the ore in a rotary furnace if it is to be eventually re-smelted in a blast furnace. The answer is simple: the sponge contains a reduced iron, free from impurities and gangue, and its re-smelting in a blast furnace requires but a low consumption of coke, many times smaller than that required for the operation of a blast furnace on common ore. Moreover, blast furnaces require rich ores, with a small content of gangue and of a specified size, otherwise they have to be concentrated and agglomerated by sintering or pelletizing.

By contrast, very fine and pulverulent ores with a high gangue content are suitable for processing in rotary furnaces. Blast furnace smelting requires a first-class fuel, coke, obtained from special grades of coal, while the direct manufacture of iron from ores may be achieved with the use of very low-grade fuels, such as coke fines and pulverized coal.

Wiberg method. Besides the rotary-furnace process, a number of other techniques are also used industrially. One of these is the Wiberg method which differs from the above described in that it employs a shaft similar to a blast furnace (Fig. 25), rather than an inclined rotary furnace.

Ore is not mixed here with the fuel, but is charged separately. The reducing mixture of carbon monoxide and hydrogen is obtained from coke in a special apparatus termed the *carburettor*. As the gases rise, they reduce iron oxides to sponge iron. The ore charged into the Wiberg furnace is heated by partially burning the gas blown into the furnace in the air injected into the top part of the furnace. Fuel consumption in the Wiberg furnace is very low.

Despite their advantages, none of the methods of direct iron manufacture can compete with the blast furnace process. The main shortcoming of these methods is their very

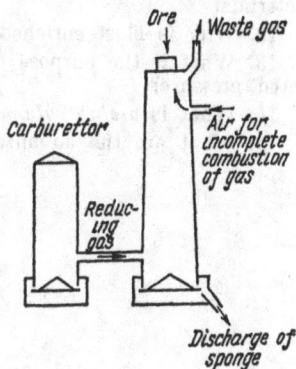

Fig. 25. Direct manufacture of iron in Wiberg furnace

low output and high cost. A large blast furnace smelts in five days the same amount of iron as the most efficient Wiberg furnace can produce in a year.

Therefore, for a long time to come, the blast furnace will remain irreplaceable as regards the manufacture of iron.

REVIEW QUESTIONS

1. Explain the fundamentals of ore agglomeration.
2. What for is natural gas blown into a blast furnace?
3. Name the chief elements of the blast furnace profile.
4. Why is blast preheated before it is injected into a blast furnace?
5. What is the difference between direct and indirect reduction?
6. Why is calcium oxide necessary for the removal of sulfur from the iron?
7. Why is the melting point of pig iron lower than that of pure iron?

8. Write the principal reactions of coke combustion in the blast furnace hearth.

9. What gases reduce the burden oxides?

10. What symptomizes the normal operation of a blast furnace?

11. Why is it necessary to remove gangue from the burden materials?

12. Why is blast enriched in oxygen?

13. What is the purpose of operating a blast furnace at an elevated pressure?

14. What is *useful volume utilization factor*?

15. What are the advantages of direct manufacture of iron?

As has been mentioned in Section II, not all machine parts can be manufactured from iron. Most of machine elements, structures, rails, bridges, ships, machine tools, vehicles, agricultural machines, etc., are made from steel.

In order to obtain a steel, the greater part of carbon, sulfur, and phosphorus should be removed down to the limits specified for various grades of steel.

Steel may be manufactured in a variety of ways. Most popular are the methods of steel manufacturing in converters and open-hearth and electric furnaces.

CHAPTER 13

Manufacture of Steel in Converters

13.1. The Bessemer Process

The design of a Bessemer converter. The converter process was the first method of large-scale manufacturing of liquid steel. It was devised and developed by Henry Bessemer, a British inventor, in 1856.

The Bessemer converter (Fig. 26) is a pear-shaped vessel made from steel sheets riveted together and lined on the inside with refractory silica firebrick. The top open part of the converter is called the *mouth*, the lower part is the *bottom* with blast *tuyères* in it.

The tuyère is a conical brick whose height is equal to the thickness of the bottom lining. Each tuyère has 9 to 12 small openings termed *nozzles* through which air is injected inside the converter. Secured to the bottom underneath is a blast box from sheet steel. Compressed air at a gauge pressure of 0.15-0.20 MN/m^2 enters this box through a side pipe

Steel Making

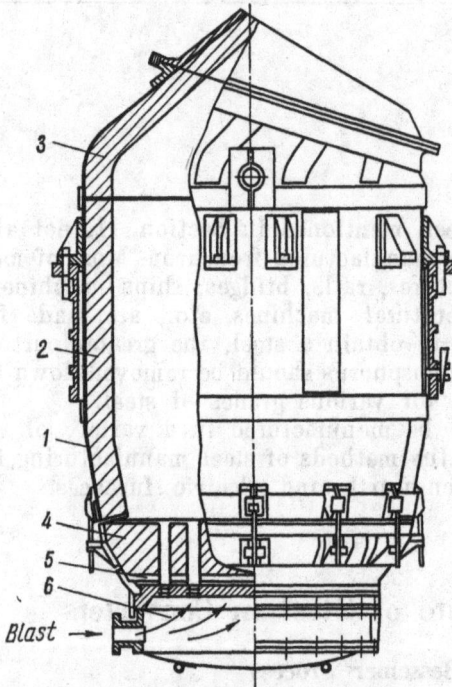

Fig. 26. Bessemer converter
1 — shell; *2* — refractory lining; *3* — mouth; *4* — bottom; *5* — blast tuyères;
6 — blast box

connected to the blower and is then distributed uniformly among the tuyères.

The bottom of the converter wears out more rapidly than the rest of the lining. Therefore, to facilitate its replacement, it is designed as a detachable assembly fastened to the converter shell with cramps and wedges.

On the outside, the converter shell carries a cast supporting ring from steel with trunnions on the sides, which serves to rotary mount the converter in the bearings of supporting frames.

One of the hollow trunnions serves to supply the blast to the converter, the other is geared to the tilting mechanism.

The nature of the Bessemer process. The Bessemer process consists essentially in blowing air through a layer of molten iron poured into the converter (Fig. 27). The oxygen of air oxidizes the silicon, manganese, and carbon of iron. In the course of blowing, the iron does not cool, rather

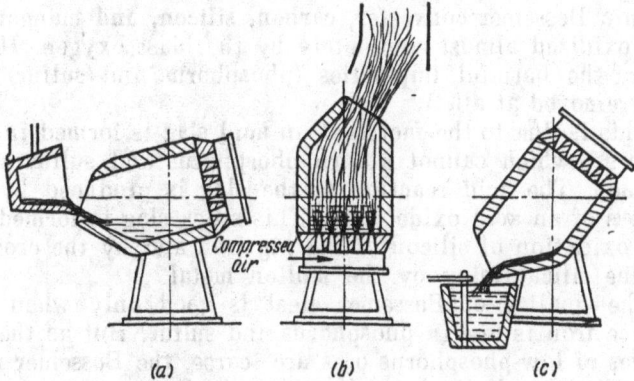

Fig. 27. Sequence of steel making in Bessemer converter
(a) pouring in pig iron; (b) blowing; (c) discharge of steel

it heats up owing to the liberation of a considerable amount of heat due to the oxidation of certain elements, chiefly silicon and carbon.

Calculations show that the oxidation of 1 kg of silicon can heat 100 kg of metal by 150-170 degrees centigrade. This is why iron intended for Bessemer processing should contain a sufficient quantity of silicon (1.0 to 1.5 per cent).

The elements in the pig iron are oxidized in the following manner. The first to oxidize are silicon and manganese, while the entire carbon remains in the melt. When silicon content in the melt decreases and the metal temperature rises, carbon burns out rapidly.

Some time (approximately 13-14 minutes for a 15-ton converter) after the elimination of silicon and manganese and burning of carbon to predetermined limits, the blow is discontinued, carbon content in the metal is checked by chemical analysis, then the converter is tilted and emptied into a pouring ladle.

Because of the high productivity and simplicity of the Bessemer process, it gained a worldwide recognition soon after its invention and found an extensive application in many countries (Britain, Russia, USA, Sweden, and others).

In a Bessemer converter, carbon, silicon, and manganese are oxidized almost completely by the blast oxygen. However, the harmful impurities (phosphorus and sulfur) are not removed at all.

This is due to the fact that an acid slag is formed in the process, which cannot absorb phosphorus and sulfur compounds. The acid reaction of the slag is produced by an excess of an acid oxide, SiO_2. This silica slag is formed by the oxidation of silicon of the pig iron and by the erosion of the silica lining by the molten metal.

The quality of Bessemer steel is good only when the source iron is low in phosphorus and sulfur. But as the reserves of low-phosphorus ores are scarce, the Bessemer process finds a limited application.

One way of removing phosphorus and sulfur is to convert them to stable chemical compounds with lime. But lime cannot be charged into a Bessemer converter, as it combines more readily with silica (silica lining included) than with the phosphoric anhydride P_2O_5 or the sulfur. Therefore, when lime is added, phosphorus and sulfur are not removed from the metal, but silica lining erodes at a great rate.

13.2. The Thomas Process

In 1878 Sidney Thomas, a British engineer, substituted a basic dolomite lining for the silica one in a Bessemer converter, and suggested the use of lime (whose chief component is the basic oxide CaO) to bind phosphorus.

The basic lining allows to obtain a basic slag (i.e., high in CaO) without sacrificing the lining. The *Thomas process* made it possible to work irons high in phosphorus.

During the blow, the phosphorus solved in the iron is oxidized by the oxygen of ferrous oxide:

$$2P + 5FeO = P_2O_5 + 5Fe$$

The phosphoric anhydride P_2O_5 is very unstable, but it gives a strong compound with lime:

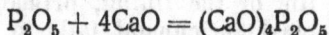

$$P_2O_5 + 4CaO = (CaO)_4P_2O_5$$

The principal operations in the Thomas and Bessemer converters are the same. The difference between them lies in that the main source of heat in the Bessemer converter is silicon, while in the Thomas process it is phosphorus whose content in the pig iron may be as high as 2.0 per cent. In the Thomas process, a high silicon content in the starting iron is undesirable.

Another feature of steel manufacture in a Thomas converter is the addition of lime in order to obtain a basic slag and remove phosphorus. In the process, the slag becomes so rich in phosphorus (18-22% P_2O_5) that it can be used as a fertilizer called phosphate slag.

The Thomas process has found wide application in France, Germany, Belgium, Luxembourg, whose iron ores contain much phosphorus. The USSR has great reserves of phosphoric ores in the Crimea.

13.3. Shortcomings Common to Thomas and Bessemer Steelmaking Processes

Simplicity of the production process and low capital investment of converter plants were the basic advantages which brought about the rapid propagation of the two methods.

However, a major shortcoming is common to both the Bessemer and the Thomas processes. The blast contains 79 per cent nitrogen which partly dissolves in the molten metal. This is why nitrogen content in finished steel may be as high as 0.020 per cent, or 4-5 times greater than that in an open-hearth steel.

Nitrogen impairs the properties of steel, especially its plasticity. In addition, such a steel is apt to "ageing", i.e., a rapid deterioration of its mechanical properties with time.

A widely used technique of nitrogen elimination from Thomas steel is the blowing of iron with oxygenated blast (30 to 40 per cent oxygen).

However, an overenrichment of the blast (in excess of 40 per cent oxygen) is undesirable because it greatly affects the service life of the lining, primarily that of the bottom. Besides, blast enrichment fails to reduce nitrogen content below 0.008-0.012 per cent.

All this puts a bar to the further development of the Bessemer and Thomas processes.

Hence, it is not surprising that the number of Thomas converters, particularly those operating on air blast, diminishes from year to year, and that very few Bessemer plants remain operative.

13.4. The Oxygen Converter Process

The advances in oxygen production in the last two decades made possible the use of commercial oxygen instead of air for blowing iron in converters. This technique is termed the *oxygen converter process*, or *basic oxygen process*, and the converters, *oxygen converters*, or *basic oxygen furnaces*.

In this process, the blast is blown not from the bottom of the converter, but from the top by means of a special water-cooled nozzle or *lance*.

Design of an oxygen converter. Let us consider the design of a 100-ton oxygen converter (Fig. 28).

The oxygen converter comprises the same principal parts as a Bessemer converter: shell *1*, bottom *2*, bearing trunnions *3*, frames *4*, and tilting mechanism *5*. However, it has a blind bottom welded to make an integral piece. The shell and the bottom are lined with refractory brick (Fig. 29); most frequently used in recent years is a lining of resin-bonded dolomite brick.

The bottom generally withstands a greater number of heats than the walls, so it is made detachable and secured to the shell by tapered bolts, which facilitates overhauls and effects a saving on refractories.

The converter shell (see Fig. 29) consists of nose section *1*, support cylinder *2* and lower truncated cone *3*. The upper part of the nose section is termed the *mouth*. The nose section is also provided with an opening for teeming steel, or *taphole 5*.

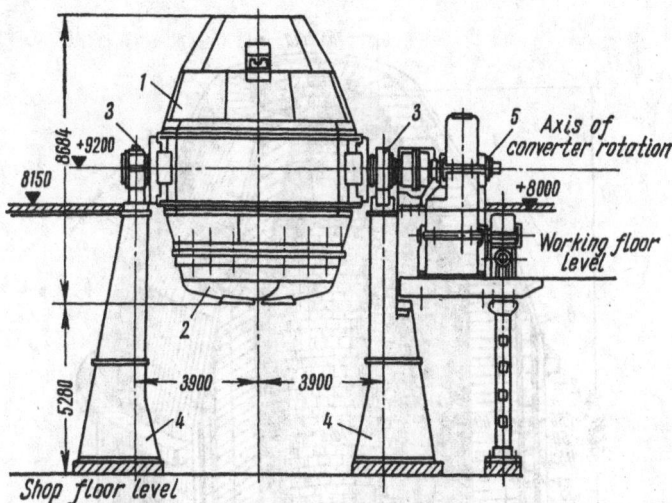

Fig. 28. Oxygen converter of 100-ton capacity

1 — shell; *2* — bottom; *3* — support bearings; *4* — frames; *5* — tilting mechanism

All parts of the converter shell are made by welding together rolled and pressed steel plates. The lower cone ends in ring *4* on which the lining rests.

The converter lining consists of a number of brick courses. Subjected to the most severe duty is the lining of the cylindrical part as this is the zone where oxygen is injected into the bath and where the reactions are very intense. Because of this the cylindrical lining is made of work layer *6* as thick as 460 mm for a 100-ton furnace, intermediate layer *7*, and reinforcement (permanent) layer *8*.

Heavy sectors for the trunnions are welded on both sides of the support cylinder. In recent designs, the trunnions are welded to a special carrier ring inside which the converter shell is secured.

The trunnions are forged pieces; the driving trunnion is geared to the tilting mechanism. The trunnions rest on rolling friction bearings mounted on the frames.

In operation the furnace must be rotated in both directions for charging the scrap, pouring the iron, tapping the

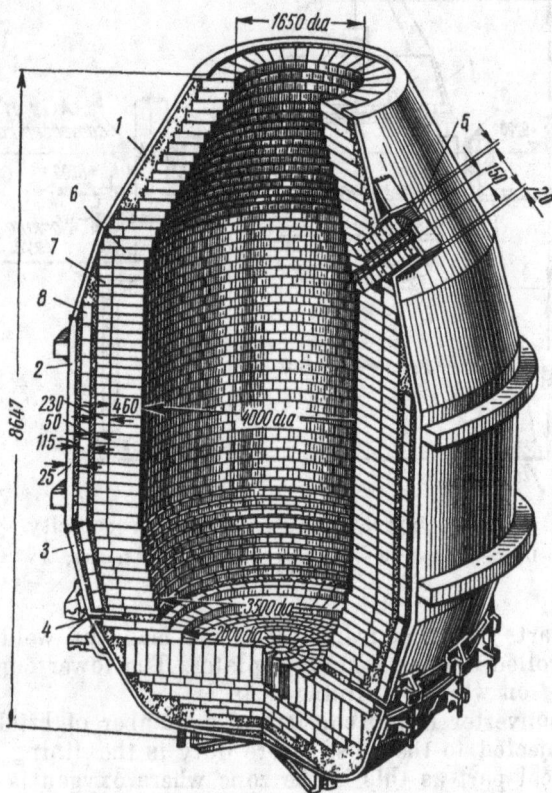

Fig. 29. Oxygen converter with lining

1 — nose section; *2* — support cylinder; *3* — lower cone; *4* — ring; *5* — taphole, *6* — working layer of lining; *7* — intermediate layer of lining; *8* — reinforcement layer

steel and slag; this is achieved with the aid of the *tilting mechanism* set on a special platform.

The tilting mechanism is rated to turn the converter through 360° in the course of 1 minute, but the speed may be reduced ten times when skimming the slag or tapping the heat.

Oxygen lance. Oxygen is blown through a special nozzle, or lance, introduced through the furnace mouth. The lance is composed of three concentric steel tubes. The central tube serves to supply the oxygen, the medium and the external tubes, for feeding and returning the cooling water.

The lower part of the lance ends in a copper tip. Up to recently tips with one large nozzle were used; but now, three- or four-nozzle tips are generally employed.

Multi-nozzle tips give better performance: the process is smoother, there are less splashes of metal and slag from the converter, and the lining lasts longer.

The process. When use is made of scrap (metallic waste of rolling-mill shops or bought-out waste), the scrap is the first to be charged (Fig. 30a). The furnace is tilted toward

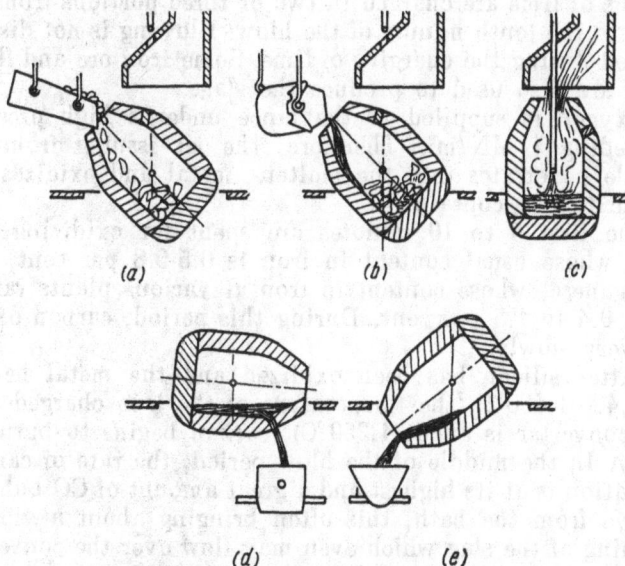

(a)   (b)   (c)

(d)   (e)

Fig. 30. Steel manufacture in an oxygen converter
(a) charging scrap; (b) pouring pig iron; (c) blowing; (d) discharging steel; (e) discharging slag

the charging aisle, and the scrap is caused to slide into the converter mouth by tilting the charging box. Some 20 to 25 tons of scrap are charged for each cycle of a 100-ton converter.

Some 100 to 110 tons of molten iron are poured directly upon charging the scrap. The ladles convey the molten iron from a mixer, a barrel-shaped vessel lined on the inside with refractories and serving for storage of molten iron. The mixer is located in a special building adjacent to the main building of the converter plant.

Once the iron has been poured, the crane with the ladle moves away, the converter is turned upright, the lance is lowered into the converter, oxygen is supplied, and the heat begins. During the blow, the tip of the lance is located at 1-1.5 m from the surface of the molten metal.

In order to form an active basic slag, lime is charged via a special pipe at the beginning of the blow. Some 7 to 9 tons of lime are charged in two or three portions from the first to the tenth minute of the blow. Blowing is not discontinued during the charging of lime. Some iron ore and fluorspar are also used to produce the slag.

Oxygen is supplied to the lance under a high pressure exceeding 1 MN/m^2. Therefore, the jet issuing from the nozzle penetrates into the molten metal and oxidizes the impurities it contains.

The first 5 to 10 minutes are spent for oxidizing silicon, whose usual content in iron is 0.5-0.8 per cent, and manganese, whose content in iron at various plants ranges from 0.4 to 1.5 per cent. During this period, carbon oxidizes very slowly.

After silicon has been oxidized and the metal heated to 1,400-1,450°C (the temperature of the iron charged into the converter is about 1,250°C), carbon begins to burn rapidly. In the middle of the blow period, the rate of carbon oxidation is at its highest and a great amount of CO bubbles evolve from the bath, this often bringing about a violent foaming of the slag which even may flow over the converter mouth.

At the end of the blow, the rate of carbon oxidation gradually decreases, and when carbon content in the metal is brought down to the required level (this is found from the character of flame, the appearance of splashes, the amount of oxygen that has been consumed, and by the duration of blow), the blow is discontinued and the lance retracted from the converter.

The blow of a 100-130-ton converter usually takes 18-23 minutes at an oxygen flow rate ranging from 250 to 300 m³/min. This time is sufficient to oxidize carbon and other impurities, to produce a basic slag, and to remove about 40 per cent sulfur and 90 per cent phosphorus. Thus, if the starting iron had 0.045% S and 0.1% P, the steel obtained from it will contain 0.025% S and 0.01% P. Nitrogen content in steel obtained by blowing with pure oxygen is 0.002-0.004 per cent, or 5 to 7 times less than that in the ordinary Bessemer and Thomas processes. Therefore, the quality of steel smelted in oxygen converters is much higher.

When the blow is over, the converter is tilted and the helpers sample the metal and the slag for a rapid analysis.

Measurement of the metal temperature. As the heat is sampled, the temperature of the metal (usually close to 1,600°C) is measured by means of an *immersion thermocouple*, whose principle of operation is as follows. When two wires of dissimilar metals are joined at ends and immersed in molten steel, the heating of the junction gives rise to an electromotive force (e.m.f.) which can be measured by a special instrument termed *potentiometer*. The higher the temperature, the greater the e.m.f. Generally used wire pairs are tungsten+molybdenum, tungsten+rhenium, as well as platinum-rhodium wires with different rhodium contents. The time necessary for taking a measurement is less than one minute.

If the chemical composition of the heat meets the specifications for a given grade of steel, the taphole is opened and metal is teemed into a ladle underneath the converter. If the chemical composition or the temperature of the metal depart from specifications, the heat may be corrected. For example, in case of a high carbon content the converter may be blown additionally. When the metal is overheated, it is cooled by adding scrap or lime.

Heat balance of an oxygen converter. As has been mentioned, the iron poured into a converter at a temperature of 1,200-1,300°C is to be heated to 1,600°C. However, when iron is blown with pure oxygen, so much heat is liberated (chiefly from the oxidation of carbon and silicon) that it is not only sufficient to heat the molten metal to 1,600°C,

but also to melt a scrap addition weighing 25 to 30 per cent
of the iron weight. It is for this purpose that scrap is charged into the converter at the beginning of a heat.

Excess heat may also be eliminated by additions of iron
ore or limestone whose decomposition absorbs considerable
amounts of heat. It should be remembered that the decomposition of 1 kg of ore or limestone consumes almost three
times as much heat as is required to melt 1 kg of scrap.
Therefore, if a heat is cooled by ore instead of by scrap,
the required quantity of ore is three times smaller than that
of scrap.

Generally, the cooling agent is scrap, whereas iron ore
and limestone are used as small corrective additions.

Deoxidation of steel. Deoxidation is the final and critical operation of steel manufacture. Employed as deoxidizers are chemical elements which combine with the oxygen
solved in the metal. These elements themselves oxidize and
their oxides float into the slag as they are lighter than the
metal. Oxygen is thus removed and the quality of steel improved.

According to the degree of deoxidation, steels are divided into rimming, killed and semi-killed grades.

Rimming steel is practically not deoxidized. In the
process of tapping, this steel is corrected for manganese
only: if manganese content in finished steel is below the
norm, some *ferromanganese* (an iron alloy with 70-80% Mn)
is charged into the ladle. On teeming, the rimming steel
effervesces (seems to boil) in ingot moulds owing to the evolution of carbon monoxide formed in the process of solidification of steel:

$$FeO + C \rightarrow CO + Fe$$

Killed steel is completely deoxidized, i.e., its oxygen
is totally combined to deoxidizing elements which have
been amply introduced in the melt. This steel does not
"boil" in the ingot mould, but solidifies calmly. Steels are
usually deoxidized by ferrosilicon, ferromanganese, and aluminium added into the ladle in amounts determined by
the operator.

Semi-killed steel is deoxidized less than killed steel,
and the character of its solidification in the ingot mould

is intermediate between those of rimming and killed steels.

Auxiliary services at an oxygen converter plant. Gases formed in the blowing of iron with oxygen have a temperature about 1,700°C and carry so much metallic dust that their cleaning is indispensable. But in order to be cleaned, the fumes must be cooled.

To accomplish this, each converter is provided with a system of flues whose walls are cooled by circulating water. The heat extracted from the outgoing gases is employed either to heat water for technical uses or to generate steam.

After passing through the uptake and the downcoming parts of the flue, the cooled fumes enter a gas cleaning system, where they are dedusted. There are a number of techniques for treating the fumes, the most widely used one being the wet method. The fumes are circulated through sections of the dust removal installations where they are sprayed lavishly with water which collects the dust and is drained off to settling tanks. Clean gases are exhausted by fans to the atmosphere through a stack.

Arranged above the converters are bins for limestone, iron ore, fluorspar and other materials. The materials are weighed, conveyed to the bins, and charged into the converters automatically.

A number of railway tracks are provided in the charging bay of the oxygen converter plant for shuttling the cars with scrap and the bogies with iron and slag ladles. The charging bay is equipped with powerful pouring cranes. A general view of an oxygen converter plant as seen from the charging bay is shown in Fig. 31.

Each oxygen converter plant is provided with a steel teeming bay equipped with teeming areas, teeming cranes, ladles, sites for repairing and drying the ladles, for assembling and drying stoppers, etc.

Oxygen converter operation. The operator controls the progress of a heat from a control desk located beside each converter or in a special gallery of the charging aisle facing the converters.

The control desk houses the devices necessary for the supervision and control of the heat: an oxygen flow rate meter; a panel stop watch indicating the time from the beginning of the blow; a device showing the distance between

Fig. 31. Converter bay

the lance tip and the metal surface; a potentiometer recording the indications of the thermocouple. Also located here are devices allowing the operator to tilt the converter for charging scrap, pouring iron, and tapping steel and slag, to raise or lower the lance, to weigh and charge the necessary materials, etc. The control desk has intercommunication facilities with various areas of the plant, the dispatching office, the mixer area, the rapid-analysis laboratory and others.

CHAPTER 14

Steel Making in Open-hearth Furnaces

The open-hearth process (or the Martin process, so named after its inventor) was first employed in 1864 and rapidly supplanted the Bessemer process owing to three main reasons:

1. Open-hearth steel is of a higher quality than Bessemer steel.

2. When the open-hearth process was devised, there were vast reserves of metallic waste which could not be utilized in the then existing converters; the open-hearth furnaces provided a means of smelting almost any quantity of scrap.

3. The open-hearth process posed no stringent requirements upon the chemical composition of pig iron.

In consequence, the open-hearth process spread rapidly and widely in the 1880's and soon outperformed the converter process by the quantity of steel produced.

14.1. Structure of Open-hearth Furnace

Consider the arrangement of a basic open-hearth furnace most widely used at the present time.

A modern open-hearth furnace consists of the following principal parts (Fig. 32):

(1) reaction chamber *10*;

(2) right and left ports composed of ports *1* proper and air uptakes *2*;

(3) air and gas slag pockets *3*;

(4) air and gas regenerators *6* with checkerwork;

(5) flues *4* for air and combustion products;

(6) a change-over system;

(7) a stack.

An open-hearth furnace is a symmetrical unit: its right and left sides are mirror images of each other.

The fuel gas and air for its combustion enter the open-hearth furnace alternatively from right and left sides. When fuel and air are supplied from the right, combustion products are exhausted to the left.

Open-hearth furnaces are provided with valves and slide gates to reverse the flow of fuel and air, this operation being called the *change-over*.

Reaction chamber is the site of steel making processes. Loaded into it through the charging doors are burden materials (metallic waste, iron ore, lime, limestone and molten iron). Once the materials are melted, the melt accumulates in the depressed part of the hearth, the *bath*.

The reaction chamber consists of the following parts: bottom *11*, front wall *14* with charging doors, and back wall *15* with built-in holes for tapping the metal and the slag.

furnace

1 — ports; 2 — air uptake; 3 — slag pocket; 4 — flues; 5 — regenerator checkerwork; 6 — regenerators; 7 — slag taphole; 8 — steel taphole; 9 — roof; 10 — reaction chamber; 11 — bottom; 12 — charging doors; 13 — charging floor; 14 — front wall; 15 — back wall;

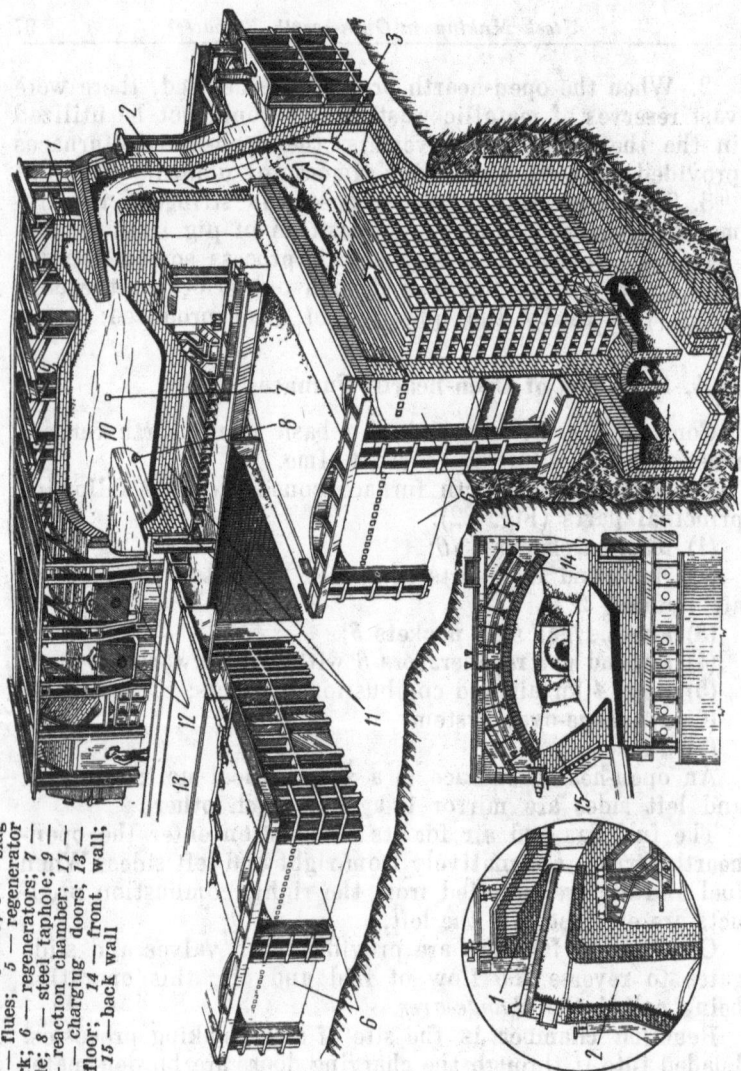

From the top the reaction chamber is spanned by roof *9*. Gas and air ports enclose the chamber at its ends.

Gas or fuel oil burns in the reaction chamber and raises the temperature there to approximately 1,800°C. The molten metal and slag wash out and erode the furnace lining. In this connection, the structure of the chamber and the refractories used should provide for an adequate service life of all its areas.

The hearth bottom made from magnesite brick is curved and slopes towards the middle, both lengthwise and across. The banks slope upwards to the walls and the furnace ends. In order to prevent the penetration of molten metal into brick-to-brick joints, the brickwork of the hearth and the banks is protected with a 150-300 mm thick blanket of magnesite powder which is fritted *in situ*.

The sloped front and back walls are made from magnesite brick, too.

Furnace roof is an essential element of the furnace, since its service life governs that of the furnace. The period of continuous operation of a furnace is the time during which no roof overhauls are carried out. This period is known as the *campaign* of the furnace, expressed by the number of heats between roof overhauls.

Roofs of modern furnaces are laid from special magnesiochromite or periclasospinelide brick. Formerly, a less resistant silica brick was used for roofs.

Roof brickwork is made up of rings (Fig. 33). Sheet iron inserts between all the bricks facilitate the interfusing of bricks in operation and add to the roof tightness. Some of the inserts project above the brick and serve to suspend the roof to metallic frames with the aid of special hangers. At the front and back walls, the roof rests upon steel skewback plates that generally are water-cooled.

A magnesiochromite roof now endures from 600 to 800 heats in small furnaces and 300 to 500 heats in large furnaces of over 400-ton capacity.

Ports and uptakes. Gas and air ports serve alternatively for supplying fuel and air to the reaction chamber and for exhausting combustion products from there.

There is a great variety of port designs. Most widely used in the Soviet Union are the Venturi-type ports (Fig. 34)

7*

Fig. 33. Roof of open-hearth furnace

which consist of one gas and two air uptakes. The air up-takes enclose laterally the gas uptake and combine, before entering the reaction chamber, into a single air conduit which supplies air to the furnace.

The ports are sloped in a manner to spread the flame over the surface of the bath. The roof of the gas conduit in the port is usually a water-cooled jacket. The cooling agent is generally water, but sometimes evaporative cooling is used.

Port refractories operate at high temperatures (approaching 1,700°C), so the brickwork is from chrome-magnesite brick. Port roofs, similarly to the furnace roof, are laid from magnesiochromite brick.

The uptakes, rectangular in cross section, direct the waste gases from the reaction chamber into slag pockets and regenerators and conduct heated gas and air to respective ports. In modern open-hearth furnaces, the uptakes are from chrome-magnesite brick.

Slag pockets. The waste gases leaving the reaction chamber carry an appreciable amount of dust composed of fine particles of slag, ore, lime, etc. In order to avoid blocking the regenerator checkerworks with dust, the regenerators are preceded by special chambers called slag pockets

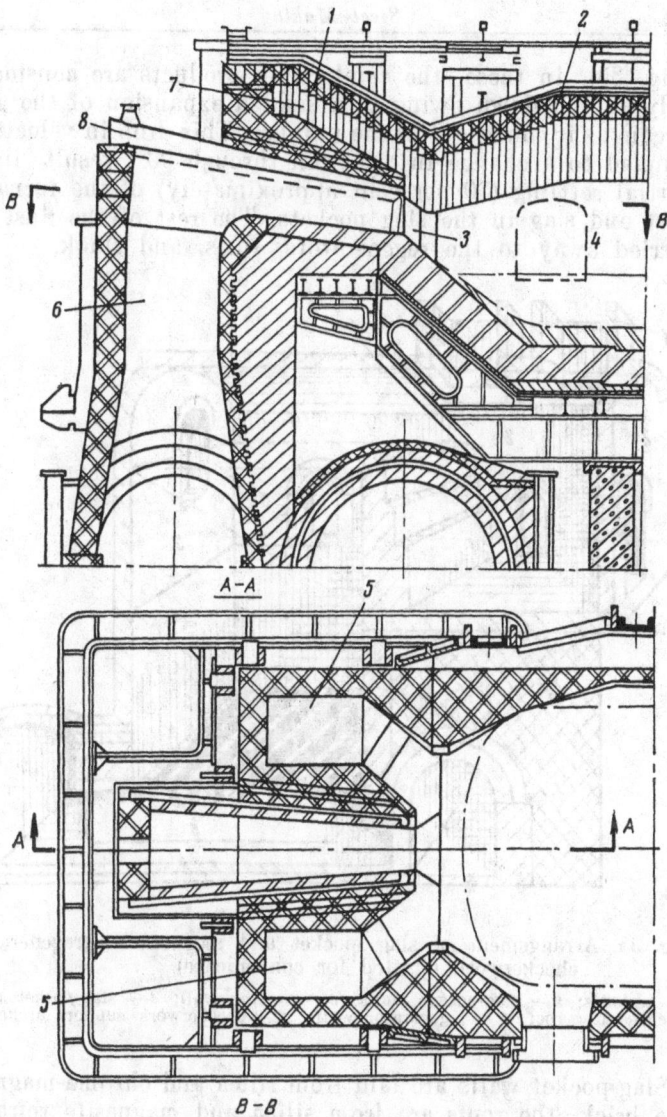

Fig. 34. Venturi-type port of open-hearth furnace

1 — port roof; 2 — main roof, 3 — premix chamber; 4 — reaction chamber; 5 — air uptake, 6 — gas uptake, 7 — port roof, 8 — water jacket in gas-port roof

(Fig. 35). In these, the combustion products are considerably slowed down owing to a sudden expansion of the gas
stream as it comes out of the uptakes. This drop in velocity,
coupled to a change in direction through 90°, results in a
partial settling (50 per cent approximately) of the furnace
dust and slag in the slag pockets. The rest of the dust is
carried away to the regenerators, flues, and stack.

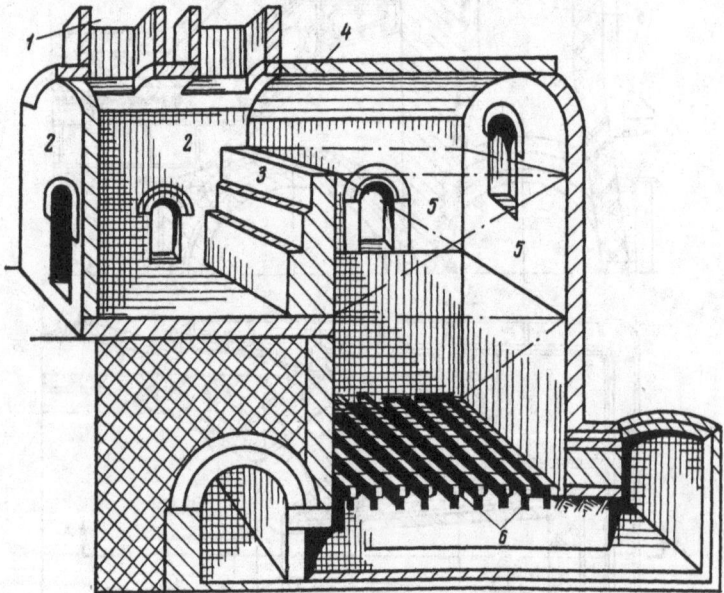

Fig. 35. Arrangement of slag pocket and regenerator (regenerator
checkerwork omitted for convenience)

1 — uptakes; *2* — slag-pocket walls; *3* — parting wall; *4* — slag-pocket and
regenerator roof, *5* — regenerator walls, *6* — checkerwork support arches

Slag-pocket walls are laid from silica and chrome-magnesite brick, the roofs are from silica and magnesite refractories.

Regenerators. The function of the regenerators (Fig. 35)
is to preheat the gas and air supplied to the furnace. Without
this, particularly when using a gas of low calorific power,

it would be impossible to obtain temperatures sufficiently high to smelt the steel in the furnace.

Air and gas are pre-heated in regenerators by the heat of the combustion products whose temperature at the slag pocket outlets is as high as 1,500°C.

As they descend through the regenerators, the hot gases give off most of their heat to the brick checkerwork supported by arches *6* (Fig. 35). When subsequently cold air or gas is passed through the hot checkerwork, it is heated to about 1,200-1,300°C.

The regenerators are composed of roof *4*, walls *5*, a bottom, a refractory checkerwork, and two chambers, one above the checkerwork and the other below it. The checkerwork is from forsterite, magnesite, fireclay or high-alumina brick. The top parts of regenerator walls are from silica, the bottom parts, from fireclay brick. The regenerator roof which is the extension of the slag-pocket roof, is laid from silica or high-alumina brick.

Flues. Flues are brick conduits intended for directing the combustion products from the regenerators to the stack and for supplying the fuel gas and air to the regenerators. The flues are generally made from fireclay brick.

Change-over arrangements serve to control the flow of the fuel gas and the air and that of the combustion products.

Each gas and air flue is provided with a valve for controlling the supply of gas and air, respectively, and with a slide gate for sealing off the flue.

Fig. 36 illustrates the operation of the change-over arrangement. The right-hand part of the figure shows that gas valve *5* is open to admit the gas to the regenerator. At the same time, slide gate *4* of the gas flue is shut down, i.e., the gas flue is isolated from the stack. The effect is that the gas from the gas main passes into the gas regenerator. In the air-supply circuit in the right-hand part of the diagram, the slide gate of the air valve is shut, while the air inlet from the fan is open, so that the air from the fan passes along the air flue to the air regenerator.

In the left-hand side of the diagram it can be seen that the slide gate of the gas valve is open, the gas main is sealed off, and the gas regenerator is connected to the stack. The same holds true for the air-supply circuit.

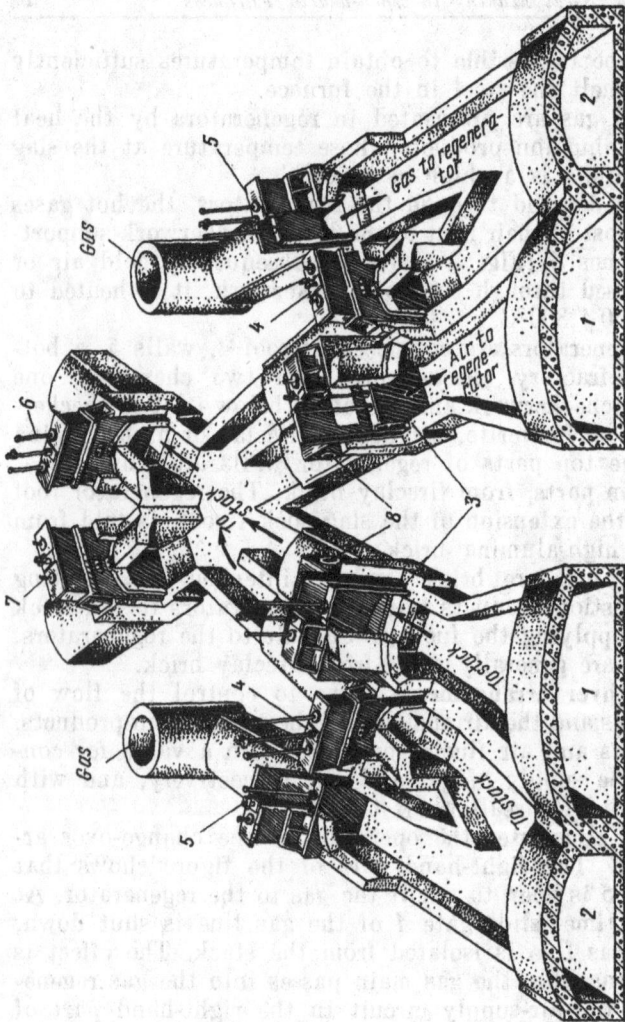

Fig. 36. Flues and change-over arrangement of gas-fired open-hearth furnace

1 — air regenerators; 2 — gas regenerators; 3 — air change-over valves; 4 — gas-flue slide-gate valve; 5 — gas valves; 6 — stack slide-gate valve; 7 — waste-heat boiler slide-gate valve; 8 — air from fan

Change-over of valves in modern open-hearth furnaces has been automated and requires no manipulation by the operator.

Stack. The stack is intended to create a natural draft. The temperature of the outgoing gases at the base of the stack ranges usually from 400 to 600°C. The density of the hot gases is less than that of the cold air, this causing a natural draft.

The magnitude of the draft depends on the height of the stack. In large-capacity furnaces, the stack may be as high as 100 m. The draft can be adjusted by the furnace operator by controlling the slide gate mounted on the main flue before the stack.

14.2. Design Features of Open-hearth Furnaces Operating on Fuel of High Calorific Value

The furnace discussed above is heated by a mixture of coke and blast furnace gases. This mixture requires preheating in regenerators to produce a high-temperature flame in the reaction chamber.

A fuel of high calorific value, e.g., fuel oil or natural gas, necessitates no preheating, this allowing a substantial simplification of the furnace design.

The construction of the furnace port is simplified since burners are used instead of the gas uptakes, and the gas circuit including the slag pockets, regenerator, and flues is omitted.

A cross section of a single-conduit furnace port for burning a gas of high calorific power is shown in Fig. 37. As the heating by natural gas requires air circuit only, the reconstruction of a furnace for operation on natural gas allows to extend the work space and so increase the furnace capacity and, consequently, its throughput.

14.3. Varieties of Open-hearth Process

Acid open-hearth process. An acid open-hearth furnace, in contrast to a basic one, is lined with silica brick.

Acid open-hearth furnaces can produce high-quality steel. However, the quality of the steel depends on the use of

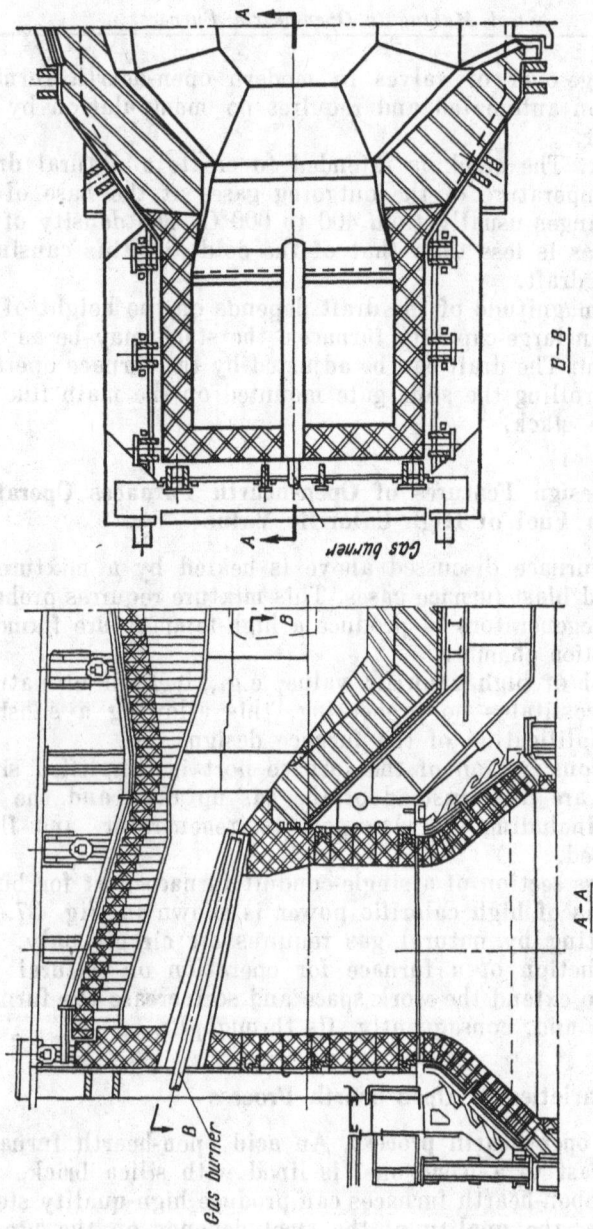

Fig. 37. Single-conduit port for furnaces fired by fuel gas of high calorific value

very pure burden materials and a fuel with a very low sulfur content.

In an acid slag, the metallic oxides are chiefly in the form of silicates $FeO \cdot SiO_2$ and $MnO \cdot SiO_2$ which hamper the removal of the harmful impurities of sulfur and phosphorus.

The solid metallic burden for an acid process is usually a semiproduct smelted in basic furnaces and containing a minimum amount of sulfur and phosphorus (less than 0.020 per cent). High-quality pig iron, produced on charcoal or low-sulfur coke, may also serve the purpose.

There are few acid furnaces in the USSR, and these have small capacities and throughput. However, because of the high quality of steel, it is these furnaces that are used to smelt high-duty steels (ball-bearing, gun steels, etc.).

Basic scrap-process. An open-hearth furnace may operate on burdens in which the proportions of pig iron and steel scrap may be most various (from 100 per cent pig iron to 100 per cent steel scrap).

The scrap-process is the conversion, in an open-hearth furnace, of a burden whose principal constituent is steel scrap. This process is usually employed only at those plants where there are no blast furnaces.

However, no open-hearth furnace operates on scrap alone, without the addition of solid pig iron or a carburizer, since the manufacture of a high-quality steel requires an effective mixing of molten metal, which is obtained by the oxidation of carbon.

This is why solid or molten iron (30 to 40 per cent of the mass of the metallic burden) is always used in the scrap-process.

Open-hearth furnaces operating by the scrap-process are generally small (up to 100 tons) and have a small throughput, and, hence, the fraction of the scrap-process in the steel output in the USSR is quite low. Almost three-fourths of all steel smelted in the USSR is produced by the scrap-ore open-hearth process. The fraction of pig iron in the metallic burden at the steel making giant plants of the Soviet Union ranges from 50 to 70 per cent.

In the discussion of the open-hearth process below we shall refer, by way of illustration, to the scrap-ore process, as it enjoys the widest application.

14.4. Fundamentals of Open-hearth Process

Fettling. The hearth bottom and banks are inevitably eroded in the process of smelting. Worn sections are restored by what is known as *fettling*, i.e., scattering over the eroded sections a refractory material, burned dolomite in lumps ranging in size from 5 to 20 mm.

The furnace is fettled either by a special fettling machine or manually. During the fettling the furnace is driven "hot", i.e., a sufficient amount of fuel is burned to ensure an adequate adhesion of dolomite lumps to the furnace lining and their fusing together. The duration of fettling is 15-20 min.

Charging of burden. In the scrap-ore process, the burden comprises a metallic portion consisting of steel scrap and molten iron and a non-metallic portion composed of iron ore and limestone.

In modern large-capacity open-hearth furnaces, the charge is comprised of 60-70 per cent molten iron and 30-40 per cent steel scrap. The quantity of the iron ore is 10-15 per cent, that of limestone is 7-9 per cent of the mass of the metallic portion of the charge.

All solid materials are conveyed to the plant by bogies in charging boxes of a capacity up to 2 m^3. Larger charging boxes (3.3 m^3) are used in newer plants.

Generally, the first to be charged is iron ore, some of which is placed on the hearth bottom. Then limestone and iron ore are spread in layers. Scrap is charged after these materials have been heated.

Solid materials are fed by charging machines.

Fuel consumption during the charging is maintained at its peak value so that the charge might be heated in the most efficient way and as rapidly as possible. Charging a large-capacity furnace with solid materials takes period of 2-2.5 hours.

Heating the burden and pouring the pig iron. When molten iron is poured on an underheated burden, the melting of materials becomes slower and the heat duration grows longer.

This is why steel scrap is additionally heated at a peak heat input once the furnace is charged.

During the heating, the charging door sill is fettled with dolomite to provide "false" sills and so prevent the slag, formed after the pig iron is poured, from splashing onto

Fig. 38. Pouring molten iron into open-hearth furnace

the charging floor. During this period, runners for pouring the iron are also set in place.

Once steel scrap has been sufficiently heated, molten iron is poured into the furnace by means of a heavy crane (Fig. 38). The iron is transported to the plant in special hot iron ladles. The pouring takes a maximum of 1 hour.

Melting. Melting is the longest single stage of smelting which takes 3-5 hours in the large-capacity open-hearth furnaces, i.e., approximately from one-third to one-fourth of the duration of a heat.

As molten iron is poured in, slag is rapidly formed of various oxides. When molten iron contacts the hot oxidized steel scrap and iron ore, impurities in the iron are oxidized vigorously by the oxygen of the scale and that of the iron ore.

The first to oxidize are silicon (whose content in pig iron ranges from 0.5 to 1.2 per cent) and manganese; these two elements react most readily with the oxygen of the scale and the ore.

Ferrous oxide FeO oxidizes silicon, manganese, and carbon as follows:

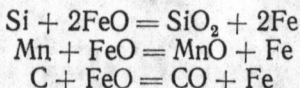

$$Si + 2FeO = SiO_2 + 2Fe$$
$$Mn + FeO = MnO + Fe$$
$$C + FeO = CO + Fe$$

Silicon and manganese oxides are poorly soluble in the metal, so they mix together to give a slag which at the beginning of the melting period consists mainly of silica SiO_2 (25-30 per cent), manganous oxide MnO (15-25 per cent) and ferrous oxide FeO (25-30 per cent). In this slag there is also a small amount of calcium oxide CaO (15-20 per cent), which passes into slag from the external layers of the limestone and the fettling materials.

A large quantity of slag in the furnace is undesirable as a thick blanket of slag hampers heat transfer to the metal and delays melting. Because of this, the largest possible quantity of slag must be removed during the first half of the melting period.

As melting goes on, the liquid portion of the metal heats up and the interaction between the iron ore and the impurities in the iron comes to an end; limestone decomposes to lime and passes into slag.

Melting is considered completed when all the solid burden materials have finally melted, no boiling is seen at the bath surface, and the slag runs freely.

Boil. The boil is the critical stage of the open-hearth process, when the metal gets rid of the remaining carbon, the harmful impurities (sulfur) are scavenged to the maximum possible degree, and the metal is heated to a temperature ensuring its normal casting into ingot moulds.

The principal reaction of the boil is the oxidation of carbon:

$$C + FeO = CO + Fe$$

Carbon monoxide CO is a gas insoluble in the metal. It forms bubbles which float up from the body of the metal and the bath seems to boil like water boils.

Gas bubbles stir the lower (cooler) and the upper (hotter) layers, and thus speed up the heating of the molten metal by the hot furnace gases.

In addition, the rising bubbles of carbon monoxide entrain some proportion of other gases (nitrogen, hydrogen) and non-metallic particles whose presence in the steel impairs its quality.

Oxidation of carbon is accelerated during the boil by small additions of iron ore or scale, while the slag composition is adjusted with lime (when the slag is fluid) or bauxite and fluorspar (when the slag is thick).

During the boil, the metal and slag are sampled to provide rapid analysis for carbon, manganese, sulfur, and phosphorus contents, and the temperature of the metal is measured by means of an immersion thermocouple.

Many furnacemen and foremen are capable of determining the carbon content in the metal by the sparks it gives when poured from a scoop and by the hardness of the sample. The lower the carbon content, the softer is the sample. The temperature of the metal may also be determined visually by observing the metal flow from the scoop, its colour and brightness.

When steel is finished, the taphole is opened and steel is tapped into a ladle. Deoxidation is performed either in the furnace or in the ladle (during tapping), depending on the grade of the steel.

14.5. Use of Oxygen for Heat Acceleration

Gaseous oxygen is used at modern iron and steel works for intensifying the combustion of fuel and for directly blowing through the metal.

To intensify the fuel combustion, oxygen is injected into the furnace at a pressure of 0.8-0.9 MN/m^2 by means of special water-cooled tuyères located in the furnace ports near the gas jackets.

Flame temperature may be increased considerably by oxygenating the blast, i.e., by partially replacing air with oxygen. This accelerates the reactions in the bath and the heating of the metal.

It is good practice to supply oxygen to the flame during the charging, heating, and melting, when maximum heat is consumed in the furnace. Oxygenation of the flame during charging and heating allows the solid burden to be charged over a shorter period of time and reduces the heating time.

The higher temperature of the flame in case of oxygenated blast during melting results in higher rates of melting, slag formation, and heating.

Practice shows that oxygenated blast raises the throughput of an open-hearth furnace by 15 to 20 per cent.

A still greater effect is obtained by blowing the molten metal with oxygen.

In the usual open-hearth process, the carbon solved in the molten metal is oxidized by the oxygen of the furnace atmosphere or of the iron ore. Oxidation of carbon by the furnace atmosphere is slow. Its acceleration by ore additions is limited by the temperature of the metal, since the heating and the decomposition of iron ore consumes much heat.

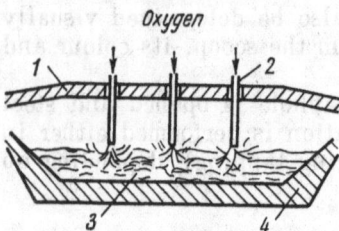

Fig. 39. Oxygen blowing of molten metal in open-hearth furnace

1 — roof; *2* — lance; *3* — molten metal; *4* — hearth bottom

By contrast, when pure oxygen is injected into the molten metal, carbon is oxidized directly by the oxygen, liberating sizeable amounts of heat. Therefore, blowing with oxygen accelerates both the oxidation of carbon and the heating of the molten metal.

Most universally oxygen is blown over the molten

bath from vertical water-cooled lances lowered toward the bath through holes in the furnace roof (Fig. 39).

A judicious combination of both methods (burning of the fuel with oxygenated blast and blowing of the molten metal with pure oxygen) may more than halve the duration of a heat.

14.6. Advantages and Limitations of the Open-hearth Process

Owing to its merits, the open-hearth process will long remain the topmost in steel manufacture.

Basic open-hearth furnaces are capable of processing iron of almost any chemical composition. The process is suited to handle any amount of low-cost steel scrap. Open-hearth furnaces can operate on any kind of fuel.

And, finally, the quality of open-hearth steel is the highest among commercial steel making processes (Bessemer, Thomas, and oxygen-converter techniques).

A limitation of the open-hearth process is a lower productivity in comparison with the oxygen-converter process. Modern open-hearth furnaces are not suited to intensification by blowing the bath with large amounts of oxygen. And lastly, the necessity to provide slag pockets and regenerators complicates, and raises costs of, the erection and running of open-hearth furnaces.

It is likely that the design of open-hearth furnaces will change gradually in the future, and leading in steel manufacture will be a type of equipment intermediate between an open-hearth furnace and an oxygen converter.

New varieties of open-hearth furnaces have now appeared. For example, there are double-bath open-hearth furnaces with two baths in the reaction chamber; in one of them burden is melted and the metal then finished by blowing with oxygen, while simultaneously solid burden is charged and heated in the other bath by burning the carbon monoxide evolving from the first bath which is being blown with oxygen. When finished steel is tapped from the one bath, molten iron is poured into the other bath; the first bath is then charged as soon as it is fettled, and so on.

Furnaces have been built with no regenerators; in this type of furnace the air for the burning of fuel is preheated in special air heaters.

CHAPTER 15

Manufacture of Steel in Electric Furnaces

The main feature which distinguishes electric furnaces from other steelmaking units is that heat is generated by electric power, rather than by the combustion of a fuel. A higher temperature can be obtained in an electric furnace than in an open-hearth furnace, this allowing the making of steels with a greater content of tungsten, molybdenum, and other hard-to-melt metals. Among the thus obtained high-alloy steels are stainless and high-speed steels.

Of major importance is the possibility of obtaining a reducing, i.e., oxygen-free, atmosphere in the hearth of an electric furnace. This reduces the burning losses of iron and readily oxidizing alloying elements that may be quite costly. A reducing atmosphere cannot be provided either in an open-hearth furnace or in an oxygen converter.

Arc-type and induction-type electric furnaces are used to smelt steel. Arc furnaces are more widely used.

15.1. Structure of Electric Arc Furnace

In an electric arc furnace, the burden materials are heated and smelted by the heat of an electric arc struck between electrodes and the metal. Fig. 40 is a schematic cross section of such a furnace.

An electric arc furnace consists of the following main parts: cylindrical shell *1*, refractory lining *2*, detachable roof *3*, two supporting frames *4*, cast bearing sectors (skids) *5*, and carbon or graphite electrodes *6*. The furnace has charging door *7* and discharge spout *8* for tapping the finished metal.

The bottoms of basic and acid arc furnaces are lined with magnesite and silica brick, respectively. The work layer of an electric furnace is rammed with a refractory mass. A basic

bottom is made from magnesite
powder bonded by resin. An acid
bottom is rammed with a mixture
of quartz sand and refractory
clay, using a small amount of
molasses as a binder.

In both cases, the mixture is
heated, applied in several layers
onto the furnace bottom brickwork,
then rammed solid, and finally
heated to a high temperature until it forms a homogeneous compound and fuses to the brick.

Walls of basic-bottom furnaces
are made of magnesite-chromite
brick set in iron boxes or of pre-
fabricated blocks made from a
mixture of magnesite powder, burned dolomite, and coal-tar pitch.

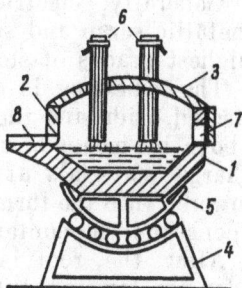

Fig. 40. Schematic cross
section of electric arc furnace

1 — shell; *2* — refractory lining; *3* — detachable roof;
4 — supporting frame; *5* —
supporting sector; *6* — electrodes; *7* — charging door;
8 — tapping spout

The roof has round holes for the electrodes. It is pre-
fabricated outside the furnace from magnesite-chromite
brick in a strong iron ring and then put on shell walls when
required. This allows a quick replacement of a worn roof.

The electrodes are secured by special electrode holders
through which electric power is supplied from a transformer via flexible cables.

The length of the electric arc and, consequently, the temperature in the furnace hearth are controlled by actuating
a special mechanism to raise or lower the electrode carriage.

Most widely used are electric furnaces with basic bottoms, since sulfur and phosphorus are readily removed in
these furnaces. The maximum capacity of basic electric
furnaces today is about 180-200 tons.

15.2. Procedure of Smelting in Basic-bottom Arc
Furnaces

The smelting of steel in a basic arc furnace differs from
that in an open-hearth furnace both in the composition of
the charge and in the procedure.

Generally, electric arc furnaces are employed to smelt metallic scrap and small amounts of solid pig iron into the highest grades of steels.

The smelting is conducted as follows. The roof is deflected aside and the burden charged onto the bottom from above by means of a bucket (the bucket is loaded with the charge materials at the storage yard). As the bucket is lowered into the furnace hearth, the lower part of the bucket opens and the contents fall down onto the bottom.

After the roof is reinstalled into position, the electrodes are lowered until they lightly contact the lumps of the burden. An arc is struck under each electrode as soon as the furnace transformer is energized, and the solid burden begins melting.

The electric-furnace process may be subdivided into the same stages as in scrap smelting in an open-hearth furnace, namely, charging, melting, boiling, and deoxidation.

As soon as the burden has melted down, some slag is run off by tilting the furnace toward the charging-door side, after first placing a slag pot under it. Most of the phosphorus is removed with the slag.

Excess carbon is oxidized during the boil. Before the boiling begins, lime is introduced into the furnace, then the bath is heated and iron ore is added.

When the carbon content has been brought to required level, ore charging is discontinued, and the furnace is kept energized for a period of time. The oxide slag is then discharged again. The slag flows over the sill of the charging door as the furnace is tilted with the electrodes raised.

Upon removal of the oxide slag, steel is refined; during this final stage, the metal is desulfurized, deoxidized, and brought to a specified chemical composition.

The metal is refined in a reducing atmosphere under a blanket of white or carbide slag, depending on the grade of steel smelted. Generally, a low-carbon steel is smelted under a white slag, and a high-carbon steel, under a carbide slag.

When a *white slag* is to be obtained, the furnace is first charged with a mixture composed of 75-80 per cent lime, 10-15 per cent fluorspar and 10-15 per cent fireclay brick waste. The mixture amounts to 2.5-3.0 per cent of the metal mass. The resulting slag is processed by a reducing slag-forming mixture (50-60 per cent lime and 10-15 per cent fluorspar, the balance being fine coke). After this, slag samples acquire a lighter hue. Finally, a mixture of ferrosilicon powder, lime, and fluorspar are added to the bath.

The white slag contains up to 65 per cent calcium oxide and only about 1 per cent ferrous oxide FeO. This slag refines the metal of sulfur and oxygen.

The *carbide slag* also contains much calcium oxide, but it differs from the white slag by the presence of calcium carbide. This slag has greater deoxidizing and desulfurizing capacities than the white slag.

While the metal is processed by the white or carbide slag, the required amounts of alloying elements are introduced into the furnace, and the metal is tapped as soon as they have gone into solution.

15.3. Manufacture of Steel in Induction Furnaces

Induction furnaces differ from arc furnaces by the method of converting electric energy into the heat necessary to smelt the metal.

The furnace (Fig. 41) consists basically of refractory cylindrical crucible *1* with coil *2* around it. The coil is composed of a certain number of turns of a water-cooled copper tubing. The coil is connected through buses to a high-frequency current generator.

When lumps of metal are charged into the crucible and the coil is energized, an electric current is induced in the metal which rapidly heats and melts the charge.

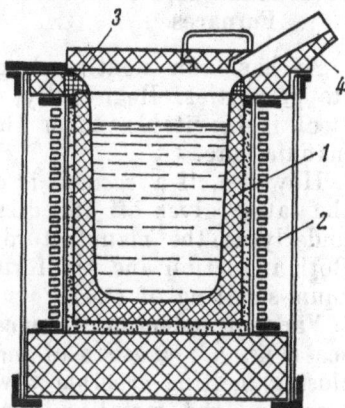

Fig. 41. Schematic cross section of induction furnace

1 — rammed crucible; *2* — coil; *3* — lid; *4* — pouring spout

The molten metal is continuously and vigorously stirred by the electromagnetic field. This stirring speeds up chemical reactions and results in a greater homogeneity of the metal. This is of a major importance for the manufacture of alloys composed of metals differing greatly in densities.

An advantage of induction furnaces is the possibility of obtaining very high temperatures throughout the body of the metal.

Induction furnaces may be used to advantage for remelting high-alloy steel scrap, as this minimizes the burning losses of costly alloying elements.

Induction furnaces are widely used for the manufacture of high-alloy low-carbon steels (0.03-0.04% C) which cannot be smelted in arc furnaces because of the carburizing action of the carbon electrodes upon the metal.

Capacities of induction furnaces are much smaller than those of the open-hearth and electric arc furnaces. Modern laboratory and industrial induction furnaces are available in capacities ranging from a few tens of kilograms to 25 tons of steel.

15.4. Manufacture of Steel in Vacuum and Electroslag Furnaces

As has been mentioned, gases dissolved in steel impair its properties. Regretfully, the penetration of gases into steel is inevitable under the conditions prevailing in its manufacture.

However, if a *vacuum* is created over the molten metal, the latter gives off the dissolved gases. This is the idea underlying the manufacture of steel in vacuum furnaces. Both induction and arc furnaces are employed for the vacuum-smelting of steel.

Vacuum induction furnaces. A vacuum induction furnace (Fig. 42) consists of melting vacuum chamber *1*, enclosing melting crucible *2* with a coil, ingot mould *3* for accepting the metal, a vacuum (suction) system, device *4* for introducing the charge, funnel *5* for admitting deoxidizers and alloying elements and measuring instruments.

First, air is evacuated from the melting chamber, then burden is charged into the crucible and the furnace is energized. Upon completion of smelting, the crucible is tilted to pour the molten metal into the ingot mould.

Smelting of steel in a vacuum induction furnace presents a number of advantages. In a conventional open-type induction furnace, the oxidation of carbon is accompanied by an increase in oxygen content. Elimination of oxygen by introducing silicon or aluminium (which serve as deoxidizers) leads to the formation of harmful non-metallic inclusions in the metal. In a vacuum furnace, gases are removed in the process of smelting, so that non-metallic inclusions are not formed.

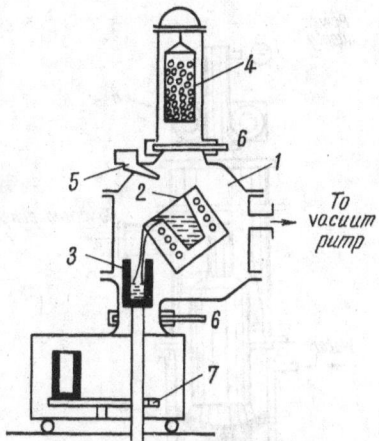

Fig. 42. Schematic diagram of vacuum induction furnace

1 — vacuum chamber; *2* — crucible with induction coil; *3* — mould; *4* — device for burden charging; *5* — funnel for charging ferroalloys; *6* — slide gates; *7* — turntable for moulds

However, the interaction between the refractory lining and the steel gives rise to some amount of inclusions. Because of this, the vacuum induction furnaces are frequently employed for the manufacture of steels for the so-called *consumable electrodes.*

Vacuum electric arc furnaces. The principle of operation of a vacuum arc furnace with a consumable electrode (Fig. 43) is as follows. As direct current is applied, an arc is struck between steel electrode *1* (which in the process serves as the charge) and starting dummy bar *2* placed on bottom plate *3* in water-cooled copper mould *4*.

The heat generated by the arc melts the consumable electrode, and droplets of molten metal pass through the arc to gradually fill the mould and form ingot *5*.

The very high temperature and vacuum created during the smelting by special pumps remove gas from the metal. The content of non-metallic inclusions is also drastically reduced.

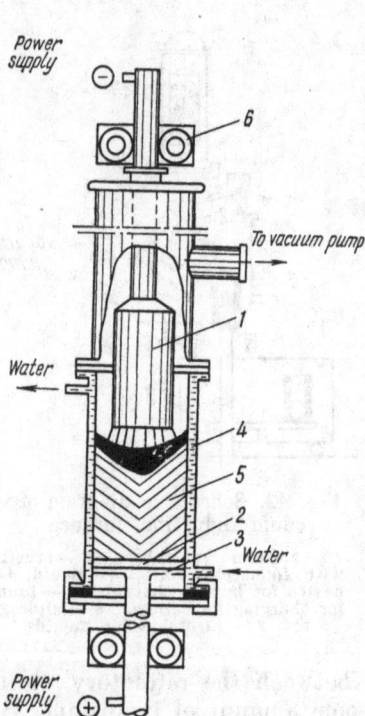

Fig. 43. Schematic diagram of vacuum electric arc furnace

1 — consumable steel electrode; *2* — dummy bar; *3* — bottom plate; *4* — water-cooled mould; *5* — ingot; *6* — electrode feeder

The crystallization of metal in the process of vacuum arc smelting differs substantially from that in the usual casting into ingot moulds, as there always is a puddle of high-temperature molten metal at the top of the ingot. The ingot solidifies from the bottom upwards. Because of this, the non-metallic inclusions float up and concentrate in the top part of the ingot. The ingot is sound, with a small shrinkage cavity and a homogeneous distribution of carbon, manganese, chromium, phosphorus, sulfur, and other elements. The quality of metal is high throughout the body of the ingot.

A limitation of the method of vacuum arc remelting is that it requires costly non-standard equipment (direct-current generators, rectifiers, etc.). In this respect, the advantage lies indisputably with the electroslag remelting of steel.

Electroslag furnaces. Electroslag furnaces differ advantageously from vacuum arc furnaces in that they require no direct current, as they operate on commercial-frequency alternating current.

The essence of the method (Fig. 44) is that a large amount of heat is liberated by passing an electric current through slag *2* which, when molten, is a resistor. This heat goes to melting the electrode *1*. Droplets of molten metal work their way through the molten slag onto the water-cooled bottom of mould 5, which shapes the ingot.

The drop-by-drop transfer of metal and the high temperature of the active slag favour the removal of gases, inclusions, and other harmful impurities.

The upward-oriented crystallization of the metallic bath also promotes the removal of non-metallic inclusions from steel and ensures the formation of a sound and homogeneous ingot structure, free from porosity, shrinkage cavities, and segregation defects.

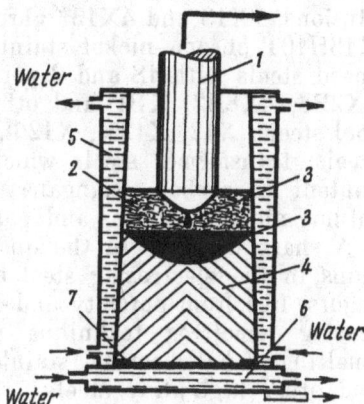

Fig. 44. Operation of electroslag furnace

1 — consumable steel electrode; *2* — slag, *3* — metal; *4* — ingot; *5* — water-cooled mould; *6* — bottom plate; *7* — dummy bar

The quality of steel manufactured by electroslag remelting depends mainly on the quality of the original material of the electrodes and correct chemical composition of the slagging flux; this flux is a mixture of fluorspar, alumina, and calcium oxide with as small content of silica, iron oxide, and magnesia as possible.

Electroslag remelting produces an extra sound metal with a minimum amount of non-metallic inclusions, characterized by high mechanical properties, particularly, plasticity and toughness at low and high temperatures.

15.5. Range of Steels and Alloys Smelted in Electric Furnaces

The reducing atmosphere and the very high temperature make the electric furnaces best adapted for the manufacture of extra pure and high-alloy steels, which cannot be smelted in open-hearth furnaces and oxygen converters.

The range of steels produced in electric furnaces includes ball-bearing steel, which must be pure of non-metallic inclusions; 1X13 and 4X13* chromium stainless steels; grade X18H10T chrome-nickel stainless steel; P18 and P9 high-speed steels with 18 and 9 per cent tungsten, respectively; 3X2B8, 4X8B2, ХВГ and other types of tungsten-alloyed tool steels; X12, X12M, X12Ф, and X12Ф1 high-chromium steels; transformer steels which require the least possible content of carbon, manganese, sulfur, phosphorus, chromium, nickel, copper, and gases, especially hydrogen.

A sharp decrease in the quantity of non-metallic inclusions in the electroslag steel and very sound structure of ingots, free from porosity and other defects, make the electroslag remelting technique particularly suited for the smelting of ball-bearing, stainless (X18H10T, 18X2H4BA), and other high-alloy steels.

CHAPTER 16

Teeming of Steel

Teeming of steel is the final stage of any steel making process. It is a critical operation which governs the final quality of steel; a steel which has been smelted to a high quality in a furnace or a converter may be spoiled by wrong teeming.

16.1. Steel Crystallization in Ingot Mould

Crystallization of killed steel. As soon as molten steel contacts the cold walls of an ingot mould, crystals are originated at many points of the rough mould surface, and solid crust *1* is formed (Fig. 45a).

The rate of crystallization decreases as the mould walls heat up; the number of points of crystal nucleation diminishes. This gives rise to *columnar crystals 2* extended towards the centre of the mould.

As the temperature of the mould walls goes up, the heat losses through them slow down, as does the rate of growth of the columnar crystals. Crystallization in the ingot centre has no definite orientation, since the crystals

---

* See designations of steels in Chapter 5, pp. 38-41.

here are formed on the rough surface of the pointed columnar crystals and on non-metallic inclusions, this giving rise to zone of *randomly oriented crystals 4.*

In the process of solidification, the volume of a molten metal decreases, which produces *shrinkage cavities 6* and

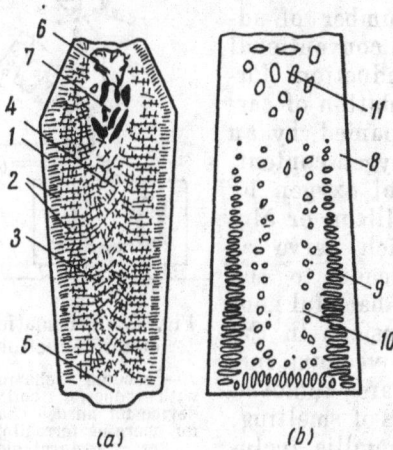

(a)   (b)

Fig. 45. Crystallization of killed (a) and rimming (b) steel

1 — fine-grained solid crust; *2* — columnar crystal zone, *3* — inclined columnar crystals; *4* — randomly oriented crystals; *5* — sound fine-grained structure; *6* — shrinkage cavity; *7* — voids and porosity; *8* — solid crust; *9* — honeycomb blowholes; *10* — secondary blowholes; *11* — gas cavities

shrinkage porosity 7. Shrinkage cavities are formed in the top part of the ingot which is the last to solidify.

The metal in the top part of the ingot must be kept liquid as long as possible. To this end the ingot top is heat-insulated by lining it with brick, covering the metal surface with special heat-insulating powders, etc. These measures result in that the shrinkage cavity is confined to the ingot top which is cut off after rolling.

Another reason for insulating the ingot head is that the non-metallic inclusions, including sulfur and phosphorus, gather in it in the process of solidification.

Crystallization of rimming steel. As has been mentioned before, rimming steel boils in the ingot mould owing

gized. Upon completion of smelting, the crucible is tilted to pour the molten metal into the ingot mould.

Smelting of steel in a vacuum induction furnace presents a number of advantages. In a conventional open-type induction furnace, the oxidation of carbon is accompanied by an increase in oxygen content. Elimination of oxygen by introducing silicon or aluminium (which serve as deoxidizers) leads to the formation of harmful non-metallic inclusions in the metal. In a vacuum furnace, gases are removed in the process of smelting, so that non-metallic inclusions are not formed.

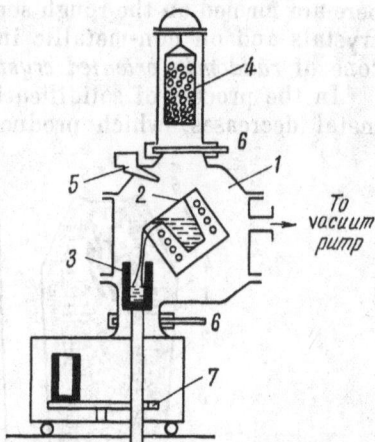

Fig. 42. Schematic diagram of vacuum induction furnace

1 — vacuum chamber; *2* — crucible with induction coil; *3* — mould; *4* — device for burden charging; *5* — funnel for charging ferroalloys; *6* — slide gates; *7* — turntable for moulds

However, the interaction between the refractory lining and the steel gives rise to some amount of inclusions. Because of this, the vacuum induction furnaces are frequently employed for the manufacture of steels for the so-called *consumable electrodes*.

Vacuum electric arc furnaces. The principle of operation of a vacuum arc furnace with a consumable electrode (Fig. 43) is as follows. As direct current is applied, an arc is struck between steel electrode *1* (which in the process serves as the charge) and starting dummy bar *2* placed on bottom plate *3* in water-cooled copper mould *4*.

The heat generated by the arc melts the consumable electrode, and droplets of molten metal pass through the arc to gradually fill the mould and form ingot *5*.

The very high temperature and vacuum created during the smelting by special pumps remove gas from the metal. The content of non-metallic inclusions is also drastically reduced.

causes additional expenses for cleaning the surface of blanks (slabs or blooms).

Bottom pouring (Fig. 46b). A number of ingot moulds are placed on a *bottom plate* of heavy iron provided with channels (*runners*). The runners are filled with hollow fire-clay brick. An ingot mould is placed at the end of each runner.

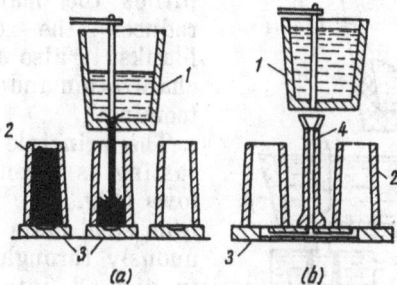

Fig. 46. Top pouring (*a*) and bottom pouring (*b*) of steel
1 — ladle; *2* — mould; *3* — bottom plate; *4* — sprue

A fireclay centre brick whose through holes are connected to the bottom-plate runners is placed at the centre of the bottom plate. A guide tube, or *sprue*, lined on the inside with special fireclay pipes is put on the centre brick.

Metal from the ladle is poured into the sprue through a trumpet and flows via the bottom plate channels simultaneously into all the ingot moulds in a gentle stream.

The surface of the bottom-poured ingots is much better than that of the top-poured ones. However, the bottom pouring has a serious shortcoming in that the molten metal partly erodes the refractory-lined runners; this results in contamination of the steel by non-metallic inclusions and impairs the properties of steel. The top-poured ingots are much cleaner.

Also, bottom pouring entails higher costs paid for an elaborate preparation of bottom plates and runners, as well as for the refractories consumed in the process.

Therefore, the choice of the teeming method for an iron

and steel plant involves a thorough consideration of the grade of steel to be poured, its application, the mass of ingots, and the operating practice.

Continuous casting of steel. Continuous casting of steel is gaining an ever wider recognition at many iron and steel plants. This is one of the signal achievements of present-day metallurgy. An efficient continuous-casting technique simplifies the manufacture and reduces the cost of steel blanks; it also allows the mechanization and automation of teeming.

The principle of continuous casting is essentially as follows (Fig. 47). Molten metal from steel ladle *1* flows continuously through intermediate tundish *2* into water-cooled copper mould *3*.

Before casting begins, a starting bar equal in cross section to the blank to be cast is introduced into the mould to serve as its bottom. As the molten metal contacts the mould bottom and walls, it begins to crystallize.

When the metal solidifies to a height of 300-400 mm above the starting bar, the bar drawing mechanism *5* is started and, on further pouring, the whole of the mould is filled with the metal.

The ingot withdrawal speed and the pouring rate are so adjusted that the metal is maintained at a constant level in the mould, and the so-

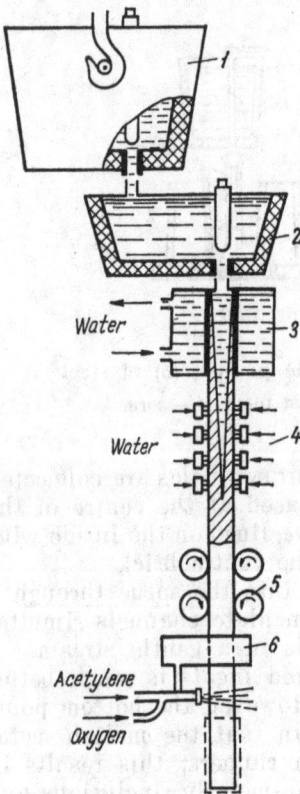

Fig. 47. Continuous casting of steel

1 — steel-pouring ladle; *2* — tundish; *3* — mould; *4* — secondary-cooling zone; *5* — withdrawal mechanism; *6* — cutting

lidifying ingot is continuously drawn out of the mould by rotating rolls.

After leaving the mould, the blank, whose core is still liquid, passes through secondary-cooling zone *4* where it is subjected to intensive cooling by atomized water, which accelerates the crystallization of the ingot core.

The cooled blank is cut by a gas torch, and standard lengths are transferred to rolling plants for further processing.

16.3. Defects of Steel Ingots

Splash. When in the process of top pouring the ladle stopper is opened very quickly, the stream of metal produces a lot of splashes which adhere to the mould walls, oxidize there, and then stick to the surface of ingots. To prevent this, the stopper should be opened in a gentle motion. There are also other measures to prevent splash, such as placing of deflecting collars from sheet steel on the mould bottom.

Lap. This defect occurs when the teeming is too slow and a solid crust formed on the metal surface in the mould folds over and is flooded by fresh portions of the rising liquid metal. Usually, this crust has time to oxidize, which prevents it from closing up during rolling, this giving rise to shells.

Shells may also be due to incorrect smelting or teeming of rimming steel, when wrong operating procedures result in the formation of a very thin, but sound external crust. When the ingots are heated prior to rolling, the crust oxidizes, the honeycomb blowholes open up, oxidize, and fail to close up in the process of rolling. The surface of rolled blanks has a "torn" appearance.

To avoid the formation of the thin crust, do not overheat the metal in the furnace or converter, keep the manganese content in steel at a reasonably low level, do not fill the moulds at excessive pouring rates, etc.

Transverse cracks may be caused by a poor inside surface of the mould (excessive wear, localized hollows) or by a gap between the mould and the hot top. In this case, the ingot may get suspended on the wall irregularities. As the crust at the outset of crystallization is still weak, this often results in a transverse crack.

Transverse cracks may be avoided by proper preparation of the moulds.

Longitudinal cracks. When the first solid layer of the ingot is very thin, the internal pressure of the molten metal may produce longitudinal cracks. This is frequently the case with round ingots of low-carbon steels.

There are a number of other defects, many of which may be corrected by chiselling the blanks or the ingots. But frequently an ingot cannot be remedied and has to be remelted.

16.4. Steel Finishing in Ladles

Vacuum degassing of steel. Gases and inclusions may be eliminated from steel not only in the furnace, but also in ladles or moulds. This may be achieved by a number of techniques, including vacuum degassing of steel.

Fig. 48*a* gives a schematic illustration of the vacuum degassing of steel in a ladle. Ladle *1* with fresh-tapped

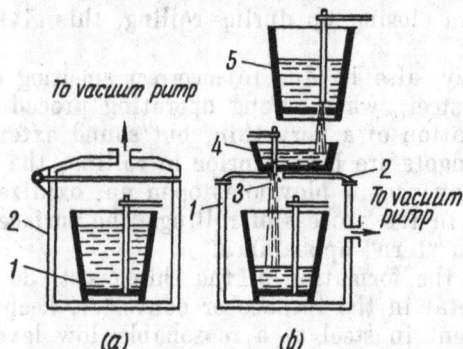

Fig. 48. Vacuum degassing of steel

(a) steel degassing in ladle   *1*—ladle with molten steel; *2*—vacuum chamber,
(b) steel degassing in stream   *1*—vacuum chamber, *2*—cover; *3*—hole in cover,
4—intermediate tundish, *5*—ladle with molten steel

steel is placed in vacuum chamber *2*. Cover *3* is put over the chamber, and the air is pumped out.

After a certain period of time, when the evolution of gases is practically over, vacuumizing is discontinued. The ladle is removed from the chamber and the steel is teemed into ingot moulds.

A shortcoming of the ladle vacuum-degassing is that steel has to be overheated in the furnace prior to vacuumizing, this causing greater wear of furnace and ladle linings and that of the stopper.

An alternative vacuum-processing technique is the vacuum degassing in stream. In this case, an empty ladle or an ingot mould (Fig. 48*b*) is placed in a special vacuum chamber. Chamber *1* is closed with cover *2* which has hole *3* tightly closed by a sheet of aluminium. Tundish *4* is placed atop the chamber.

As soon as the air has been pumped out of the chamber, steel is poured from usual ladle *5* into the tundish; the stream issuing from the latter pierces the aluminium sheet and falls down into a ladle placed inside the chamber. But since the chamber is at a vacuum, gases are continuously removed from the metal in the stream.

This technique is frequently employed at machine-building plants for vacuum degassing heavy ingots weighing 100 tons or more.

Processing of molten steel by slags. One of the methods for improving the properties of steel in a ladle is to process the molten steel by special slags.

To do so, a mixture of lime and alumina is melted in a slag-melting furnace. The liquid slag which has a great capacity for binding sulfur is poured in required amounts into the ladle wherein the metal is to be tapped from an open-hearth furnace or a converter.

The metal falls from the furnace into the ladle in a strong stream, this resulting in vigorous stirring of the metal and the lime-alumina slag. The intimate contact between the metal and the slag liberates the metal from sulfur and non-metallic inclusions.

CHAPTER 17

Performance of Steelmaking Units

Of the total quantity of steel manufactured in the USSR, some 80 per cent are smelted in open-hearth furnaces.

The chief performance factor of an open-hearth furnace is its *yearly output*. This factor allows a comprehensive evaluation of furnace operation, as it takes into account all the idle periods that occurred over the yearly period.

The yearly output of open-hearth furnaces tends to increase due to bringing down the furnace idle time, improving the service life of the refractory lining, accelerating the heats by a more efficient use of oxygen, improving the production planning, mechanizing and automating various operations.

Another major factor of performance is the *fuel consumption*. This factor is greatly affected by the "hot" idle time of the furnace. Reduction of the idle time, an efficient management of the furnace crew, and elimination of operational delays are the means to cut down the fuel consumption.

The fuel consumption can be drastically reduced if use is made of various intensifiers, such as oxygen or compressed air, the gain being the largest when they are blown right through the molten metal.

Of major interest is the *yield of finished steel*, which is the ratio of the mass of marketable ingots to the mass of the metallic burden (iron, metallic scrap, deoxidizers, iron in iron ore).

The lower the teeming losses, the losses due to furnace failures, and the losses of iron with slag, the greater the amount of iron passing to finished marketable ingots and the higher the yield of finished steel.

An important economic factor is the *cost of steel*, i.e., the expenses for the manufacture of 1 ton of marketable steel. This cost is the sum of the costs of molten iron, scrap, iron ore, limestone, lime, deoxidizers, fuel, oxygen, electric power, water, and refractory materials consumed in the smelting of 1 ton of steel. Therefore, an economical use of the materials and fuel, and careful operation of the furnace with the view to prolonging the service life of refractories contribute to a lower cost of steel.

In recent years, steel is being manufactured on an ever increasing scale in oxygen converters which smelt the metal at a faster rate than any other steelmaking unit. For example, it takes only 40-50 minutes to process a charge in a 100-ton oxygen converter, whereas the duration of a heat in a 400-ton open-hearth furnace is 9 to 11 hours.

An oxygen-converter plant comprising three 100-ton converters has a yearly output in excess of 2 million tons of steel. The same output at an open-hearth plant would require 5 or 6 furnaces of a capacity of 400 tons each.

Obviously, three oxygen converters cost less than six open-hearth furnaces with the heavier and more expensive crane handling equipment. Besides, the oxygen converters require no fuel, so they are more efficient than the open-hearth furnaces.

CHAPTER 18

Mechanization and Automation of Steelmaking Processes

18.1. Mechanization

Most of the production process operations in open-hearth furnace, oxygen-converter, and electric-furnace plants are mechanized. The amount of manual work is decreasing steadily.

The burden (iron scrap, ore, limestone, and other materials) is loaded into charging boxes by electric magnet or clamshell cranes and fed to the furnaces and converters by special charging machines or cranes.

Widely employed are machines for fettling and guniting (an operation similar to fettling), and for demolishing worn lining in oxygen converters and steel landles.

Special devices for internal facing of ingot mould walls under the control of a single operator are successfully employed in the preparation of moulds.

Mechanization is continually being introduced in furnace-repair operations. Widely used during open-hearth overhauls are belt conveyors for loading broken bricks and other waste into boxes or railway cars and various types of winches for cleaning regenerator checkerwork and flues.

Slag pockets are cleaned with the aid of various machines which load the loose slag into special boxes. Of great help in furnace repair are teeming and charging cranes, lift trucks, etc.

Oxygen-converter plants are equipped with special machines which facilitate the work of operators. Carriage-mounted jacks of a high lifting capacity are employed to remove or install the furnace bottoms. Special telescopic elevators are employed to aid in lining the oxygen converters.

Many mechanisms at modern oxygen-converter plants are remote-controlled, i.e., operated from a distance.

For example, the bogie with a steel ladle is remote-controlled by an operator stationed in a special cabin where he is exposed to no hazard. A similar arrangement is used to control a self-propelled bogie which transports charging boxes in the charging bay of oxygen-converter plants.

In all of the modern oxygen-converter plants, loose materials (lime, ore, fluorspar) are weighed and charged into the converter automatically.

In steel-making plants, samples of metal are conveyed by compressed air along special pipes. This relieves the furnacemen from the necessity of carrying the samples to the laboratory and speeds up considerably the analyses of samples.

Efficient communication facilities between various work areas in steel-making plants are provided by extensive telephone networks, while a number of plants, particularly the oxygen-converter plants, are equipped with intercommunicating networks. Radio and television are also employed for the purposes of control and operation.

18.2. Automation

Much attention is being given to the automation of steel-making processes. Many operations which control the rate of heat supply to open-hearth furnaces are automated.

The instruments which record the temperature of the regenerator checkerwork, are connected to the valve change-over systems. As soon as a predetermined temperature of the checkerwork is reached, the valves are automatically changed over, this maintaining the normal heating of gas or air and improving the service life of the checkerwork.

At some plants, the automatic valve change-over is time-controlled. To achieve this, the system controlling the frequency of change-overs is connected to timers which automatically maintain a preset time interval between the change-overs.

The instruments which record the pressure in the furnace are linked to the flue slide-gate mechanisms, this allowing to maintain automatically the required pressure in the furnace.

Techniques for measuring the temperature of molten metal in the furnace have been developed in recent years. Application of these instruments will introduce further refinements into the automatic control of smelting processes.

There are good prospects for automating the oxygen-converter process. Up to now, only individual operations have been automated, but it may be expected that the great advances in computer engineering and the introduction of techniques for rapid determining the chemical composition of metal and slag would provide a firm basis for a complete automation of the oxygen converters in not so distant future.

If all the necessary data (mass of molten iron and steel scrap; chemical composition and temperature of iron; grade of steel, and other information) are fed to a computer, it will quickly calculate the amount of lime and iron ore to be charged into the converter and the moment when the blowing is to be discontinued.

REVIEW QUESTIONS

1. What is the essence of the Bessemer method of steel making?
2. What is the difference between the Thomas and Bessemer steel-making methods?
3. Name the shortcomings common to methods of steel making in bottom-blown converters.
4. What are the principal parts of an oxygen converter?
5. What is the basic difference between the Bessemer and Thomas methods and the oxygen-converter process?

6. How is steel smelted in an oxygen converter?
7. How is the temperature of molten steel measured?
8. What is the deoxidation of steel and how is it carried out?
9. Describe the essence of the open-hearth process.
10. Name the principal elements of an open-hearth furnace and describe their functions.
11. Name and describe the stages of the open-hearth process.
12. What are the impurities of iron that are oxidized in the process of smelting?
13. Write the principal reactions taking place in an open-hearth furnace during the burden melting.
14. Describe the features of the boil stage in the open-hearth process. What methods are used for controlling the smelting procedure?
15. Why does the application of gaseous oxygen accelerate the open-hearth process?
16. Describe the design of an electric arc furnace.
17. Name the principal stages of smelting in an electric arc furnace.
18. What are the advantages of steel manufacture in induction furnaces?
19. What are the features of steel manufacture in vacuum-arc and induction furnaces?
20. What is the difference in the crystallization of killed and rimming steels?
21. Describe the advantages and shortcomings of top and bottom pouring.
22. What are the features of continuous casting of steel?
23. Name the defects possible in a steel ingot.
24. Is it possible to improve the quality of steel after it has been tapped from the furnace or converter?
25. What are the performance characteristics of a steel-making unit?

Castings constitute more than 50 per cent of the total number of pieces used in machines. Many pieces of the machines used at iron and steel plants are also cast. At the present state of the art, even the most intricate types of machine parts can be produced by casting.

The basic knowledge of the casting process and the terminology used in foundry practice can be obtained by

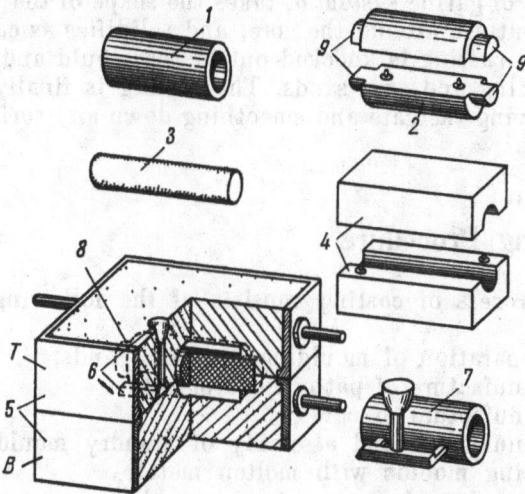

Fig. 49. Moulding of bushing

1 — bushing; *2* — pattern; *3* — core; *4* — core box; *5* — foundry mould; *6* — gating system; *7* — casting; *8* — supporting seats; *9* — core prints

referring to Fig. 49 which shows schematically the moulding of a simplest part, bushing *1*.

In order to make a mould with a cavity corresponding in dimensions and shape to the bushing, pattern *2* must be manufactured from wood or metal, depending on the number of parts to be produced. The pattern is split along its axis of symmetry to facilitate moulding, and is thus composed of two halves. The through hole in the bushing is formed by core *3* made from core sand in core box *4*. The core is longer than the bushing by the length of the core prints on which the core rests in the mould. The pattern is provided with core prints *9* to produce supporting seats *8* in the mould.

Foundry mould *5* is composed of the top (*T*) and the bottom (*B*) halves. The mould halves are manufactured separately by ramming and packing the sand in two frames of steel or cast iron, called flasks.

Molten metal is poured into the mould cavity through channels of gating system *6*, takes the shape of the void left by the pattern around the core, and solidifies as casting *7*. Then the casting is knocked out of the mould and cleaned of moulding and core sands. The casting is finally fettled by removing the gate and smoothing down any surface irregularities.

CHAPTER 19

Moulding Procedure

The process of casting consists of the following operations:
(1) preparation of moulding and core sands;
(2) manufacture of pattern equipment;
(3) manufacture of cores;
(4) manufacture and assembly of foundry moulds;
(5) filling moulds with molten metal;
(6) knock-out of castings from moulds;
(7) cleaning of castings;
(8) heat treatment of castings (if necessary).

19.1. Types of Foundry Moulds

Foundry moulds may be subdivided into the following types:
(1) expendable moulds intended for the manufacture of a single casting and destroyed when the casting is extracted;

(2) semi-permanent moulds used for the manufacture of a few castings;

(3) permanent moulds used for the manufacture of many castings.

Expendable moulds are manufactured from moulding sand, i.e., a mixture of sand, clay, and special binders. They may be used wet (green) or dryed before being filled with metal.

Classed as expendable are also the so-called shell moulds and precision lost-wax moulds.

Semi-permanent moulds are manufactured from special refractory sands, then burned at high temperatures. A careful knocking out of the castings allows the semi-permanent moulds to be reused tens of times.

Permanent moulds are made from cast iron or, less frequently, from steel. They are rated to manufacture hundreds and thousands of moderate-size castings.

Castings are mostly manufactured in expendable moulds. Semi-permanent moulds are employed chiefly for large castings of relatively simple shape (ingot moulds and slag pots); permanent moulds are employed for batch casting.

19.2. Moulding and Core Sands

Foundry moulds and cores are mostly manufactured from mixtures of natural sands and clays with the addition of the necessary amount of water. The composition and properties of the materials and mixtures depend on their expected service in the mould.

Moulding and core sands must meet the following requirements:

(1) mechanical strength sufficient to enable the mould to withstand assembly, transportation, and the impact of a stream of metal;

(2) gas permeability, i.e., the ability to pass air and other gases evolving from the mould in the process of pouring and solidification of the metal;

(3) refractoriness: on contact with molten metal the sand should not melt, soften, frit or adhere to the casting;

(4) pliability, so that the sand should not obstruct the shrinkage of the metal as the latter solidifies and cools;

(5) heat conductivity, affecting the rate of cooling of the metal in the mould as it solidifies;

(6) long service life, which permits the sands to retain their properties after repeated use;

(7) low cost.

The moulding sands are classified by the content of silica and that of the bonding clay (Table 4).

Table 4

Moulding Sands

Name of sand	Content of bonding clay, per cent	Content of silica, per cent
Quartz	2	90 - 97
Quartz-fluorspar	2	90
Weak	2 to 10	—
Medium strong	10 to 20	—
Strong	20 to 30	—
Extra strong	30 to 50	—

Clays are classed by refractoriness in the following manner: highly refractory (not less than 1,580°C), medium refractory (not less than 1,350°C); low refractory clays are not standardized.

Adhesion of the moulding sand to the metal is minimized by the addition of 5 to 15 per cent pulverized coal to clay and sand. For iron castings, the quantity of pulverized coal is the greater, the larger the casting. When dry moulds are employed, the burning of the moulding sand to the casting may be prevented by facing the moulds with a thin coat of black wash composed of graphite, coke, silica flour, and talc.

The surface of green moulds is faced with parting powders. Used as the parting powder are pulverized coal, graphite, or cement.

Oven-dried moulds are made from sands with admixtures of fine sawdust, straw, peat, and other substances that burn in the process of drying, form small pores, and thus increase the gas permeability of the sand. Mechanical strength of the moulding sands is enhanced by sulfite lye, water glass, and other binders. The rate of casting solidification is controlled by introducing magnesite, chromite, and other materials into the moulding sands.

Used as special binders which facilitate the knock-out of cores from casting are vegetable and mineral oils, synthetic resins, rosin, bitumen, food molasses, dextrin and pectic glue.

Core sands should have higher physical properties than moulding sands. The core sands should also be easily removable from castings. The cores are subjected to heavy duty and are utilized dry. Before the finished moulding and core sands are cleared for use, they are checked for moisture content, gas permeability, and mechanical strength.

The quality of moulding and core sands greatly affects the quality of castings, therefore, much care is given to the proper preparation of core sand.

Quartz and clay sands are subject to drying and screening to separate foreign matter (rocks, pebble, wood splinters) which may contaminate it during quarrying and transportation. Clay used as a powder is dried, crushed, ground, and screened. Coal is dried and pulverized.

Foundry moulds are frequently made from reclaimed (used) sands. The sand is knocked out of the moulds, crushed (for lumps from dry moulds), passed through magnetic separators to remove metallic inclusions (splashes, nails, sprues), screened, and cooled to 30-35°C.

The thus prepared materials are mixed dry in predetermined proportions, moistened, and mixed again to uniform composition ensuring the required properties. The sands are then matured in bins to obtain an even distribution of the moisture and clay. Maturing takes 2-3 hours. Prior to use, the sand is fluffed to enhance its gas permeability.

In preparation of moulding sands use is made of mechanical screens, mixing and crushing roll mills, aerators, pug mills, and other machines which facilitate work and raise its efficiency.

19.3. Pattern Equipment

Sand moulds are manufactured with the aid of wooden or metallic patterns whose external shapes coincide with those of the castings. Pattern equipment includes also the core boxes which serve to produce the inside cavities and holes in the cast parts, as well as the gate system patterns, pattern plates, mould boards gauges, etc. The pattern dimensions are somewhat larger than those of the casting to allow for the shrinkage (bodies expand when heated and contract, or shrink, when cooled). Shrinkage values for different alloys are given in Table 5.

Table 5

Shrinkage Values of Common Cast Alloys

Alloy	Shrinkage, per cent	Alloy	Shrinkage, per cent
Cast iron	0.5 - 1.0	Brass	1.0 - 1.5
White cast iron	1.5 - 2.0	Aluminium	0.8 - 1.1
Carbon steel	1.5 - 2.0	alloys	
Manganese steel	2.8 - 3.0	(with silicon)	
Tin bronze	1.0 - 1.5	Magnesium alloys	1.2 - 1.4

The design of the patterns should provide for their easy removal out of the moulding sand without damaging the moulds. To achieve this, intricate patterns are split into parts that can be easily joined together by dowels.

The mechanical strength of wooden models depends on the wood used (limewood, pine, maple, alder, and others) and on the manufacturing procedure.

A stronger pattern is glued from separate parts with due regard for fibre orientation, rather than made from a single piece of wood.

When designing and manufacturing the patterns, account should be taken of the moulding conditions that may affect the wear resistance of patterns: rammer impacts in the process of sand ramming; impacts in jar ramming and squeezing machines; stresses arizing in the pattern when it is removed from a mould, and when the cores are removed from the pattern; friction of pattern surfaces on contact with the solid particles of the moulding materials; and a great many other factors.

In order to identify the patterns intended for castings of different materials, they are painted in different colours. In machine moulding for mass production, the patterns are made from metallic alloys (grey iron, aluminium, and copper), as well as from gypsum and cement.

For precision casting, the patterns are made from low-melting materials, such as a mixture of stearin and paraffin.

19.4. Manufacture of Cores and Moulds

Manufacture of cores. When a hollow casting is desired, a core is placed inside the mould to prevent it from being filled entirely with molten metal. In the resulting casting, the hollow part has the shape and dimensions of the core.

When making the cores, shrinkage of the metal is also taken into account. Linear dimensions and volume of the cores should be larger than the dimensions of the cavity in the castings by the value of the shrinkage of the metal.

As has been mentioned above, cores are made from clay, sand, and waste moulding sand with admixture of binders. Cores are oven-dried for 5-10 hours at 200-400°C, depending on the core material composition and the properties of the binder. In the process of drying, the moisture vaporizes and the binder sets, providing the required mechanical strength of the cores.

Cores are manufactured manually in core boxes or by machines.

A plain cylindrical core (Fig. 50) is made in a box composed of two halves, which are aligned by dowels and secured by a clamp. Core sand is put inside the box and rammed with the box set vertically. When the box has been filled to approximately half its height, the core grid is placed at the centre and ramming is carried on. Upon filling with sand, a few through holes are pierced with a wire to allow the gases to escape, after which the clamp is removed, the two halves of the box are opened, and the finished core is allowed to dry.

Moulding machines, sand slingers, as well as special extrusion, core-blowing, and core shooter machines are em-

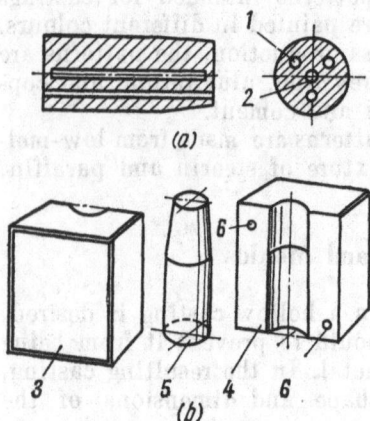

Fig. 50. Core manufacture

(a) cross sections through core; (b) core box;
1 — vents; 2 — grid wire; 3 — cope; 4 — drag; 5 — core, 6 — dowels

ployed to produce cores on a mass scale.

The cores are located and secured in the moulds with the aid of special supporting extensions, or core prints, and with supports termed chaplets.

Manufacture of moulds. Moulds are made in a variety of ways. Plain and large parts, such as plates and grates, are manufactured by pouring the metal into moulds formed in the foundry floor.

The standard practice is moulding in flasks, or moulding boxes, which are rigid frames from steel or cast iron (Fig. 51) or, sometimes, from wood.

The mould for casting a plate (Fig. 52) may serve as an illustration of a coreless mould.

Pattern *1* is laid on mould board *2* made of dry wood with the surface smoothly planed. The length and the width of the mould board should be somewhat greater than those of the box.

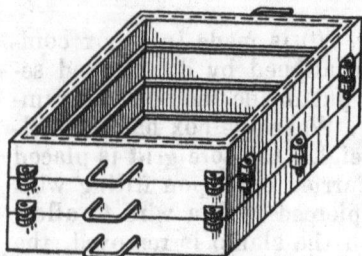

Fig. 51. Moulding box

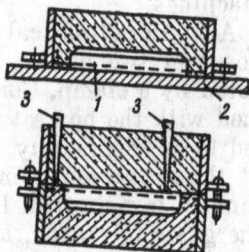

Fig. 52. Box moulding

1 — pattern;   2 — mould board;   3 — wooden cones

The pattern and the drag box are put on the mould board with the box enclosing the pattern. The pattern is then powdered with dry quartz sand, the box is filled to capacity with moulding sand, then it is rammed, and excess sand is struck off by a straight-edge. The box with the mould board is turned over, the board is removed and the cope box is put atop the drag box. The cope box is also filled with the moulding sand, and two wooden cones *3* are inserted in it to form two cavities, of which one serves as the gate, or the channel for pouring the metal, and the other is the vent through which gases escape from the mould.

The boxes are then separated gently, the pattern is removed, the boxes are reconnected to produce a finished mould for casting a plate.

Machine moulding is an effective means to raise the productivity of labour, provide greater accuracy and interchangeability of castings, effect a saving on metal (10-15 per cent) and bring down the total cost of castings.

At present, moulding sands are transported by belt conveyors to special bins, from which the required amount of sand is drawn in a matter of seconds into a moulding box by opening the metering device.

Moulding machines are distinguished by the method employed to ram the moulding sand, by the method of pattern removal from the moulds, and by the drive of the machine.

By the method of sand ramming, the moulding machines are classed as hand-operated, jolt, and squeeze-moulding machines, and sand slingers. The squeeze-moulding machines may have top or bottom sand frame. Most widely used are the jolt moulding machines.

Fig. 53 illustrates schematically the ramming of moulding sand by jolting.

Table *1* of the jolt machine receives mould board *2* which supports mould box *3* filled with sand from the bin. Under the action of compressed air entering through hole *4*, the table of the machine rises together with the sand-filled box, and then falls down, hitting a solid rest block. This jolting action compacts the sand in the flask. Compressed air is exhausted through hole *5*.

According to the method of pattern removal from the mould, the moulding machines may be divided into pin-

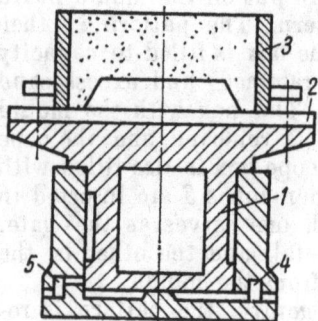

Fig. 53. Jolt moulding machine

1 — table; *2* — mould board; *3* — mould box; *4* — air supply; *5* — air exhaust

lift, stripping-plate, turntable, and roll-over moulding machines.

As to the drive, hand-driven, hydraulic, pneumatic, and mechanical moulding machines are known.

In recent years, an advanced casting technique involving the use of moulds from chemically setting mixtures has been introduced industrially, and self-setting free-running foam-type moulding mixtures are being successfully tested on an industrial scale.

Casting in chemically setting moulds is specific in that the mixtures containing soluble glass set rapidly when processed by carbon dioxide. When moulds are manufactured from chemically setting mixtures, the vaporization of moisture is replaced by a chemical process in which the water and silicon dioxide composing the soluble glass form a stable compound (silica gel). The film of the gel binds the sand particles into a strong mass, thus ensuring the mechanical strength of the mould working layer. In foundry practice, a wide variety of high-quality castings from cast iron, steel, and non-ferrous alloys in sizes from a few kilograms to tens of tons are produced in thin-wall, thick-wall, and shell moulds made from chemically setting mixtures.

Soluble-glass mixtures set in the air of the shop, since the soluble glass absorbs carbon dioxide to form the silica gel. Therefore, ambient temperature should not exceed 10°C. The moulds should be held in the air for 3 to 25 hours, depending on their mass.

Advantages inherent in the use of free-running self-setting mixtures in foundry work are as follows: low cost and good availability of the materials for preparing the mixtures; simplicity of manufacture; setting within a few minutes with neither drying nor carbon dioxide processing being necessary; replacement of a labour-consuming ramming of moulds and cores by simply pouring the mixture into the boxes; higher accuracy of castings; easy knock-out of moulds and cores; possibility of mechanizing and automating the casting process.

The refinement of the long-established processes and the introduction of new techniques run parallel with continual mechanization and automation of production processes. Auxiliary devices of moulding machines (distributing bins, feeding metering devices, positioners for assembling moulds, etc.) are being refined, existing machines are being modernized, new types of machines are being designed, and automated moulding lines are being created.

An automated moulding line of a foundry shop may be visualized as follows.

The moulding boxes are filled with exact portions of sand with the aid of a numerically controlled machine. Sand-filled boxes are transported by an automatic conveyor to a special automated installation where the sand is squeezed under a high pressure and the boxes are subjected to a high-frequency vibration (no such operations are required for the boxes filled with free-running mixtures). The assembly of the mould, including the superposition of the cope box upon the drag box, their accurate alignment and clamping, is mechanized.

The core department is also automatically controlled. The cores are manufactured from free-running self-setting or chemically-setting mixtures. Exact portions of the core mixture are fed under the control of a computer. Bulky drying ovens are replaced by compact installations for blowing the cores with carbon dioxide, the installations being part of the continuous core-manufacturing line.

CHAPTER 20

Casting Alloys and Their Properties

20.1. Casting Properties of Alloys

The castability of an alloy depends on its high fluidity, low shrinkage, and moderate segregation.

Fluidity is the property of a melt to fill the thinnest sections of a mould.

The fluidity of iron and steel increases with an increase in the content of carbon, silicon, and especially phosphorus. Sulfur and chromium impair fluidity, while the effect of manganese and nickel is negligible.

The castability of an alloy is affected, besides its chemical composition, by the degree of overheating prior to pouring and by the temperature of the mould.

Shrinkage is a decrease in the volume of an alloy as it passes from liquid to solid state. Shrinkage depends upon the chemical composition of the alloy, its rate of cooling, and the temperature of pouring. High shrinkage causes internal stresses in castings, which may give rise to the formation of cracks. Besides, excessive shrinkage results in large cavities and porosity at places that are the last to solidify.

The values of shrinkage for the most common foundry alloys are given in Table 5. Shrinkage of iron castings is the smaller, the higher is the graphitization of the iron. Pouring in dry moulds and a greater thickness of casting walls lessen shrinkage.

Segregation is heterogeneity of the chemical composition at different points of a casting.

Sulfur, phosphorus, and carbon show a clear tendency to segregate in iron and steel castings. In large-size steel castings, the difference in the content of these elements, as compared to the mean analysis of the liquid metal, reaches 500 per cent for sulfur and 300 per cent for phosphorus and carbon, this causing different mechanical strength at various points of castings.

In addition to the above properties, a casting alloy should have a moderate melting point and a low gas-absorbing ability, i.e., a small capacity for dissolving gases when molten, so that a sound casting could be obtained.

20.2. Gray Iron and Its Melting

Cast iron is the most widely employed casting material. It is cheaper than any other foundry material, has good castability, and is sufficiently strong for the manufacture of many machine parts. The principal properties of gray iron (fluidity, shrinkage, and segregation) are the most favourable ones, and gray iron serves as the standard for comparing the casting properties of other alloys.

The quality of iron castings depends, in the first place, on the correct selection of charge materials. The charge is composed of pig-iron ingots with admixtures of iron scrap, waste steel, and fluxes. The fluxes are commonly limestone and basic open-hearth slag.

The fuel for cupola are coal coke (replaced sometimes by anthracite or thermo-anthracite) and natural gas.

The charge is so chosen as to obtain a cast iron of the desired chemical composition.

Cast iron castings should have specified hardness and strength in tension and bending. These parameters are basic for classification of iron castings (Table 6).

Table 6

Mechanical Properties of Some Grades of Soviet-Made Gray Iron

Grade of cast iron	Tensile strength, MN/m^2	Bending strength, MN/m^2	Brinnel hardness
СЧ 12 - 28	120	280	143 - 229
СЧ 15 - 32	150	320	163 - 229
СЧ 18 - 36	180	360	170 - 229
СЧ 21 - 40	210	400	170 - 241
СЧ 24 - 44	240	440	170 - 241
СЧ 28 - 48	280	480	170 - 241
СЧ 32 - 52	320	520	170 - 241
СЧ 35 - 56	350	560	170 - 241
СЧ 38 - 60	380	600	197 - 262

The letters СЧ stand for gray iron. The first two digits in the grade designation denote the tensile strength, the next two, the bending strength.

Mechanical properties of gray iron depend mainly on the form of the graphite and on its quantity in the structure of iron, and, to a smaller degree, on the structure of the bulk of metal.

A comparison of the mechanical properties of cast irons shows that pearlitic cast iron has a higher strength whatsver the form of graphite, while ferritic cast iron is more

10*

plastic, especially in the forms with spheroidal (globular) graphite.

High-quality iron castings are obtained by the addition of sizable amounts of steel scrap or special additives, as well as by an adjustment of the structure of iron through heat treatment, inoculation, and alloying.

Castings from alloy cast iron (with admixtures of nickel, chromium, molybdenum, titanium, vanadium, copper, etc.) possess high strength, resistance to corrosion, acids, alkali, and high temperatures.

A cast iron is inoculated by introducing a small amount (0.1-0.5 per cent of the iron mass) of special additions (calcium silicon, ferrosilicon, alsifer), which promote the formation of structures with fine-grained graphite and pearlite, this improving the mechanical properties of cast irons.

Successful inoculation of cast iron requires that the temperature of metal tapped from the cupola be no less than 1,420 °C, otherwise the effect of the additives is appreciably weakened.

Inoculation with magnesium results in a high-strength cast iron with spheroidal graphite. It is good practice to use the high-strength graphite for castings of intricate shapes with abrupt changes in cross section, such as crankshafts, gear wheels, pump housing, etc. The high-strength cast irons are designated by the letters ВЧ; the first two digits stand for tensile strength, the next two, for percentage elongation.

Cast iron is melted in cupolas, stationary and tilting reverberatory furnaces, and electric furnaces.

Cupola design and operation. A cupola is the melting unit most widely used in foundry shops. It is simple in design and servicing, consumes less fuel than any other metal-melting furnace, can heat cast iron to high temperatures, and possesses a high throughput capacity.

The cupola (Fig. 54) is a typical shaft furnace. Its shaft consists of welded or riveted steel shell *1* lined with fireclay brick *2*. Bottom *3* is rammed with moulding sand to form hearth *4*, the ramming being done through cleaning door *5* which is then bricked up. Compressed air for burning the fuel is supplied from circular wind box *6* through tuyères *7*. The burden materials are introduced through charging hole

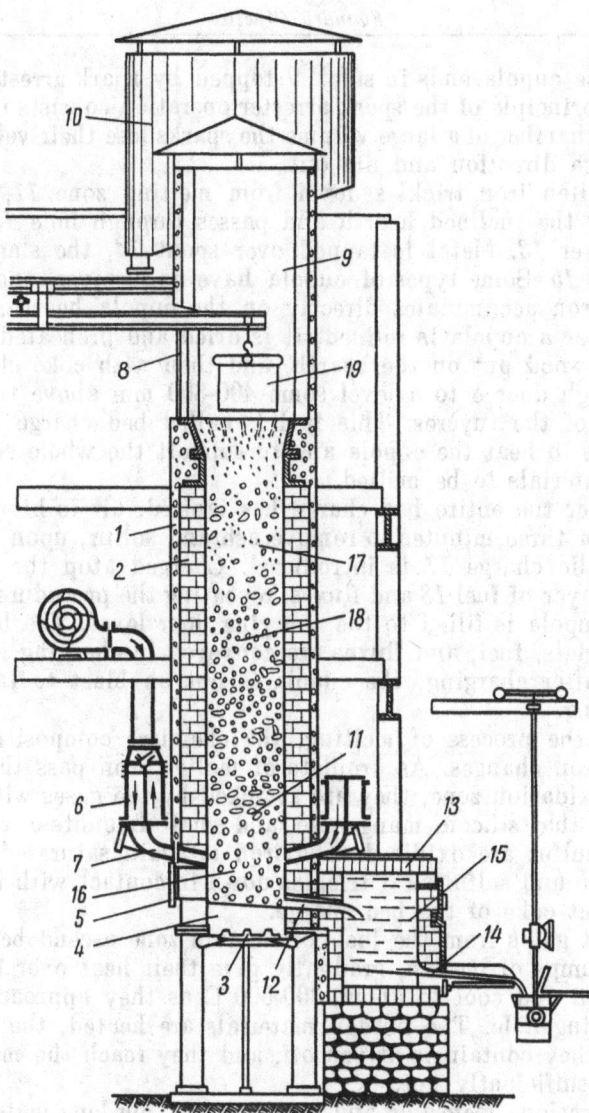

Fig. 54. Cupola furnace

1 — shell; 2 — refractory lining; 3 — bottom; 4 — hearth; 5 — cleaning door;
6 — wind box; 7 — tuyère; 8 — charging door; 9 — stack; 10 — spark arrester;
11 — melting zone; 12 — hole; 13 — receiver; 14 — iron spout; 15 — slag
spout; 16 — bed charge; 17 — metallic charge; 18 — melt charge; 19 — bucket
for burden and fuel charging

8. The cupola ends in stack *9* topped by spark arrester *10*. The principle of the spark arrester operation consists in that in a chamber of a large volume the sparks lose their velocity, change direction and die out.

Molten iron trickles down from melting zone *11*, flows along the inclined hearth and passes through hole *12* into receiver *13*. Metal is tapped over spout *14*, the slag over spout *15*. Some types of cupola have no receiver, and molten iron accumulates directly on the cupola hearth.

After a cupola is relined, it is dried and preheated, first with wood put on the hearth, and then with coke charged through door *8* to a level some 400-800 mm above the top edge of the tuyères. This fuel is called bed charge *16*. It serves to heat the cupola and to support the whole column of materials to be melted.

After the entire bed charge has ignited, air is blown for two or three minutes to remove ash and sulfur, upon which metallic charge *17* is introduced. Charged atop the latter is a layer of fuel *18* and fluxes, repeating the procedure until the cupola is filled to the charging door level. The burden materials, fuel, and fluxes are conveyed in charging bucket *19*. After charging, the cupola is put on blast to initiate melting.

In the process of melting, the chemical composition of the iron changes. As droplets of molten iron pass through the oxidation zone, they are attacked by the gases with the effect that silicon, manganese, and small amounts of carbon and sulfur are oxidized. The iron is again saturated with carbon and sulfur as it trickles down in contact with incandescent coke of the bed charge.

Hot gases from the fuel-combustion zone ascend between the lumps of burden, gradually give their heat over to the burden and cool to about 300-500°C as they approach the charging hole. The burden materials are heated, the moisture they contain is driven off, and they reach the melting zone sufficiently hot.

Selection, metering and weighing of burden materials, lifting them to the charging platform and charging them into the cupola can be mechanized and automated.

20.3. Malleable Iron Casting

Malleable cast iron is obtained by heat treatment of white iron. This name is a conventionality, as malleable iron is not subjected to any working which made use of its malleability.

Malleabilizing of iron castings consists in annealing the castings in neutral or oxidizing atmosphere in such a way that cementite is transformed to iron and temper carbon (graphitization) or cementite is decomposed and the surface carbon is eliminated. The temper carbon formed in the case of annealing in neutral medium makes casting fractures have a black velvety appearance (black-heart malleable cast iron); after annealing in an oxidizing medium (iron ore), the castings acquire a white fracture (white-heart malleable cast iron).

The two varieties of malleable cast iron differ in microstructure, mechanical properties, and chemical composition.

Malleable cast iron is used to cast pieces with a mean wall thickness of 5 to 20 mm. If necessary, parts can be made as thin as one or two millimetres or as thick as 50 mm. However, from the economic standpoint, it is preferable to produce thick-walled castings from steel or spheroidal-graphite iron.

Malleable cast iron differs from gray iron in that the temper carbon (in the black-heart malleable cast iron) is rounded rather than flaky as in gray iron. This makes the mechanical properties of malleable cast iron superior to those of gray iron.

The present trend is to increase the manufacture of items from black-heart malleable iron, as they require shorter annealing procedures than those from the white-heart iron.

Black-heart malleable cast iron can be obtained with either ferritic or pearlitic base metal. Before annealing, the fracture of castings should be white.

When choosing the chemical composition of a cast iron, it is advisable to reduce, as far as possible, the total carbon content and to raise the silicon content in the white iron. This will reduce the time of annealing.

An example of chemical composition of a black-heart malleable cast iron is given in Table 7.

Table 7

Chemical Composition, Per Cent, of Malleable Cast Iron

Element	Before annealing	After annealing
Total carbon including:	2.2 - 2.9	1.8 - 2.9
graphite carbon	≤0.02	1 6 - 2.7
fixed carbon	2.2 - 2.9	0 1 - 0.2
Silicon	0.8 - 1.4	0.8 - 1.4
Manganese	0.3 - 0.5	0.3 - 0.5
Phosphorus	≤0.20	≤0.20
Sulfur	0.05 - 0.15	0.05 - 0.15

An accelerated annealing for black-heart malleable cast iron in a chamber-type furnace takes some 40 hours, after which the castings at about 650°C are discharged into the air. Moulding and core sands used for the manufacture of malleable iron castings are the same as for castings from gray iron. The shrinkage of white iron is approximately 50 per cent larger than that of gray iron. This should be taken into account in the manufacture of patterns.

Castings from malleable iron are widely used in all branches of industry, serving as small parts for motorcars, tractors, farm and textile machinery; small fittings are also made of malleable cast iron.

20.4. Steel Castings

Steel castings are produced in sizes from a few kilograms to several hundreds of tons. They are of major importance to heavy engineering, such as the transport, road-building, metallurgical, or machine-building industries.

Steel is inferior in casting properties to cast iron, and its melting point is considerably higher, it has a poorer fluidity and a greater shrinkage. However, steel castings excel cast-iron ones in mechanical properties, particularly in plasticity and notch toughness.

Carbon steel intended for foundry work should contain not more than 0.5% C, not more than 0.37% Si, up to 0.8% Mn and a minimum amount of phosphorus and sulfur.

Table 8

Mechanical Properties of Carbon-Steel Castings

Steel	Tensile strength, MN/m²	Percentage elongation	Notch toughness, MN m/m²
15 Л	200	24	50
20 Л	220	23	50
25 Л	240	19	40
30 Л	260	17	35
35 Л	280	15	35
40 Л	300	14	30
45 Л	320	12	30
50 Л	340	10	25
55 Л	350	11	25

Mechanical properties of carbon steels used for castings are given in Table 8.

The letter Л in the grade designation indicates that the steel is intended for foundry work, the digits show the mean carbon content in hundredths of 1 per cent. Casting steels are conventionally subdivided into a number of groups depending on the percentage of alloying additions (chromium, nickel, molybdenum, tungsten, copper, and titanium); in a carbon low-alloy steel, the content of alloying additions does not exceed 2.5 per cent; a medium-alloy steel may have from 2.5 to 10 per cent of additions, and a high-alloy steel, in excess of 10 per cent.

Steel castings are usually annealed or normalized to relieve internal stresses.

Casting steels are melted in side-blown converters, small open-hearth furnaces, and electric furnaces.

CHAPTER 21

Special Methods of Casting

Modern engineering places upon castings such stringent requirements as to their strength, dimensional accuracy, and surface finish that cannot be satisfied by parts cast in sand moulds. These requirements can be met by special casting techniques.

Gravity-die casting. As has been mentioned above, permanent moulds made from cast iron or steel are termed gravity dies.

Gravity dies for steel casting withstand some 600-700 cycles of work, while those for cast iron parts of plain shapes can produce several thousand castings.

Gravity-die castings have high mechanical properties, uniform fine-grained structure, high dimensional accuracy, good finish, and often are used as-cast, without any additional cleaning or machining. Gravity-die casting raises the productivity of labour 2.5 to 3.0 times as compared to casting into expendable moulds.

The service life of gravity dies and the quality of castings may be improved by facing the die walls with special refractory compounds, such as whitening or carbon black, the latter being applied by a smoking flame of acetylene.

Gravity-die castings may be mass-produced by special casting machines of advanced design.

Centrifugal casting. This method is based on the fact that in a rapidly rotating mould a molten metal is thrown off the mould surface by the centrifugal forces and takes the mould configuration on freezing. Centrifugal casting is used for the manufacture of pipes and other cylindrical items.

Solidification of metal under the action of centrifugal forces favours considerable compacting and raises the strength of the item, since all the gases, non-metallic inclusions, and other impurities are segregated by the centrifugal forces and appear at the internal surface of the hollow casting.

In the centrifugal method the castings may be rotated about a horizontal axis or about a vertical axis (Fig. 55). Rotation of the mould about a horizontal axis has a wider application. Water and sewer pipes, shells for piston rings (ring pots), gears, and other items are cast in horizontal centrifugal machines (Fig. 55a). Vertical machines are used to cast parts of a small height, such as wheels and pulleys (Fig. 55b). The speed of the mould rotation in horizontal centrifugal machines ranges from 40 to 80 rad/s, in vertical machines, from 200 to 300 rad/s.

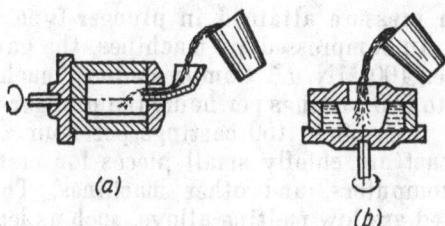

Fig. 55. Centrifugal casting

(a) horizontal machine; (b) vertical machine

Pressure-die casting. The method consists essentially in that molten metal fills steel moulds under the pressure of a plunger or compressed air and then solidifies; the castings are practically ready for use after the sprue is cut off.

Fig. 56 presents a schematic diagram of a vertical-plunger cold-chamber machine. On the left-hand sketch, molten metal is poured into injection chamber *1*. On the middle sketch, the metal is forced by injection plunger *2* into die *3*. On the right-hand sketch, the mould opens, the injection plunger rises, and closing plunger *4* rises to cut off the remainder of the sprue and push it out of the injection chamber.

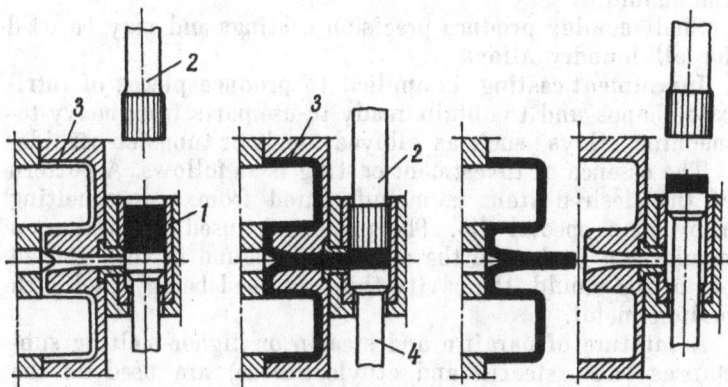

Fig. 56. Pressure-die casting

1 — injection chamber; *2* — injection plunger; *3* — die; *4* — closing plunger

Maximum pressure attained in plunger-type machines is 200 MN/m^2. In compressed-air machines, the gauge pressure of air reaches 100 MN/m^2. Compressed-air machines are capable of 50 to 500 castings per hour, the plunger-type machines produce from 50 to 150 castings per hour.

Pressure-cast are chiefly small pieces for motorcars, motorcycles, computers, and other machines. The materials generally used are low melting alloys, such as lead-tin, zinc, aluminium, and magnesium alloys.

Shell-mould casting. The method is essentially as follows. A metallic mould board together with the patterns it carries is heated to 180-200°C; then the patterns are faced with a moulding compound composed of sand, special synthetic resin, and soluble glass. When heated, the compound transforms into a non-fusible and insoluble substance.

The layer of the compound adjacent to the heated plate and pattern sinters to make a strong shell 10-12 mm thick. Non-sintered compound is removed, the mould board with the sintered shell is charged into a furnace at temperatures of 250 to 300°C, where the shell acquires a still greater strength. The half-mould thus obtained is joined with another half-mould manufactured in the same way. The complete mould is installed in a box, the empty part of the box is filled with sand, and molten metal is poured into the mould.

Shell moulds produce precision castings and may be used for all foundry alloys.

Investment casting is applied to produce pieces of intricate shapes and to obtain ready-to-use parts from heavy-to-machine alloys, such as alloyed steels or tungsten carbide.

The essence of investment casting is as follows. A pattern of the desired item is manufactured from a low-melting alloy in a special die. The pattern is used to produce a mould in a flask, and the pattern compound is then melted out of the mould, the cavity thus obtained being filled with molten metal.

A mixture of paraffin and stearin or higher-melting substances (e.g., stearin and ethylcellulose) are used as the pattern materials.

The sequence of operations in investment casting is as follows:

(1) manufacture of a master pattern which serves to make the patterns;

(2) manufacture of a die for producing the investment patterns;

(3) preparation of the fusible compound for the patterns;

(4) manufacture of a pattern in the die;

(5) preparation of the mould and fusing out the pattern;

(6) baking of the mould;

(7) melting the metal for the casting, filling the mould with metal and leaving the casting to cool;

(8) knocking out, dressing, and fettling the casting.

Moulding for investment castings is different from the usual moulding procedure. Most important here is the facing layer which must be applied so as to ensure both the dimensional accuracy of the casting and the strength of the mould. The facing layer is usually obtained by immersing the pattern into a bath filled with a suspension of quartz flour and hydrolyzed ethyl silicate (100 g of quartz flour per 45-55 ml of ethyl silicate).

The pattern is plunged into the suspension 3 to 6 times. After the first immersion, the pattern (or a set of patterns) is blown with sand and treated with ammonia in a drawer-type drier during 5 to 10 minutes. After holding the mould in the air for 10-40 minutes, the pattern is immersed again, and so on. The first immersion is into a fresh suspension, after which used suspension may be employed.

The coated pattern is then placed into a moulding box and the voids between the pattern and the box walls are charged with a dry or wet filler.

Upon drying, the pattern material is melted out of the mould, the moulds are baked, and conveyed to the pouring area.

Investment castings feature a high dimensional accuracy and require no machining. Their mass may range from 1 to 100 kg, the walls may be as thin as 0.3 mm.

CHAPTER 22

Casting Defects, Their Prevention and Remedy

There are many casting defects, the most important ones being cavities, cracks, burn-ons, misruns, etc.

Cavities are usually found when the casting is machined. They are classified as shrinkage cavities, blowholes, sand holes, and slag inclusions.

Shrinkage cavities and porosity are depressions and voids of irregular shape, appearing at the places that are the last to solidify in casting.

The size, shape, and location of a shrinkage cavity are governed by the shrinkage properties of the metal involved, as well as by the size of the casting, method of pouring, temperature of the poured metal, its rate of cooling in the mould, etc.

As has been shown, shrinkage cavities may be prevented by a thermal insulation of the riser and gradual topping up of the riser and removal of the riser when fettling the casting.

Shrinkage porosity, i.e., a dense concentration of small fissures and voids in the body of casting, may appear at various points of castings, commonly near the shrinkage cavities.

The porosity is detected by pumping water at a high pressure into a casting.

Blowholes are spherical or egg-shaped voids with the surface smooth and brilliant in blowholes closed inside the casting or oxidized in the blowholes on the casting outside.

They are due to the gases dissolved in molten metal and evolving as the casting cools; the bubbles of gas tend to float up, but they cannot rise in the freezing metal and thus are trapped inside the casting.

Blowholes caused by a poor quality of the metal are most often small in size and dispersed throughout the body of the casting.

Blowholes may also occur because of defective moulds and wrong pouring: poor gas permeability of the moulding sand coupled with an intensive evolution of gases, high rate of pouring, interrupted pouring, incorrect design of the gating system, etc.

Sand holes are internal or external irregularly shaped cavities in the casting, filled partially or entirely with the moulding material; the cavities may be due to the mould erosion by the stream of metal, damage to the mould during the withdrawal of the pattern, or use of poor moulding materials.

Slag inclusions are chiefly caused by slag penetration in the mould as the metal is poured. To prevent slag from getting into the mould, the gating systems must be provided with dirt traps. However, even with the proper design of the gating system, the slag may still penetrate the mould if the gating system is not filled adequately with metal throughout the pouring or if pouring is interrupted.

Cracks may be hot and cold, through and blind ones. *Hot cracks* are formed in the process of metal cooling and are due to internal stresses arising because of excessive shrinkage and non-uniform cooldown at different points of the casting.

Cold cracks are discontinuities of metal appearing at the end of solidification as a result of internal stresses caused by shrinkage.

As the hot cracks appear at high temperatures, their surface is always oxidized, while in cold cracks it is clean or covered with light temper colours.

Burn-ons may be due to chemical (burn-on proper) or mechanical (metal penetration) causes. Poor refractoriness of the moulding sand favours a chemical bonding of the sand to the casting accompanied by the formation of easy-melting compounds of the sand with the oxides of iron, manganese, etc., and these compounds may penetrate deep into the casting. This type of burn-on is very difficult to remove, it requires pneumatic chiseling or grinding.

The metal penetration is chiefly due to a high porosity of the facing sand, a high pouring temperature of the metal, and an excessive pressure (or head) of the metal, when filling high casting moulds.

Misruns occur when a portion of the casting is not filled with metal because of its poor fluidity, trapping of gases or vapours in the mould, or leakage of the metal through the mould joints.

Cold laps are casting defects similar in shape to cracks, but with rounded edges. They occur when a mould is filled with an insufficiently fluid metal or when pouring is interrupted.

Foundry defects cause heavy losses to the national economy.

Defects may be prevented only if their causes have been detected and corrective measures devised.

Defects may be minimized by taking the following measures:

1. Thorough inspection of the starting materials (burden materials, moulding sands, etc.).

2. Choice of correct production procedures (moulding, melting, pouring, etc.).

3. Strict compliance with production procedures and organizational discipline.

4. Proper quality inspection of castings.

5. Introduction of efficient methods for correcting casting defects.

Today, many foundries include special repair shops where casting defects are remedied.

Leaks in castings, detected by hydraulic testing, are corrected by pressure-impregnation with bakelite varnish (followed by heat treatment at 150-180°C) or by welding of the leaky points.

Castings may be gas- or arc-welded after the defective portion of the metal is removed.

Shrinkage defects and misruns in large-size castings (such as ingots) are corrected by the so-called "liquid" welding, which consists in pouring molten iron into a properly-cleaned defective portion of the casting.

REVIEW QUESTIONS

1. What is the essence of casting?
2. What properties are required of moulding and core sands?
3. What materials go into the making of foundry moulds?
4. From what materials are patterns fabricated? What requirements must be met by these materials?
5. What kinds of moulding procedures are there? What is the essence of machine moulding?
6. What properties are essential to casting alloys?
7. How are iron castings graded?
8. What is the cupola operating principle?
9. What is the difference between iron and steel castings?
10. What is gravity die and centrifugal casting?
11. What operations are involved in pressure-casting?
12. Describe the procedure of casting in shell moulds.
13. Speak on the procedure of investment casting.
14. What defects may occur in castings?

SECTION V • PLASTIC WORKING OF METALS

Finished articles may be obtained from metals and alloys by a variety of methods: casting, rolling, or machining. Over three quarters of the metal, smelted in the USSR, are subjected to various kinds of plastic working, chiefly by rolling.

The principal kinds of plastic working of metals are rolling, forging, stamping, pressing, and drawing. The goal of each of the techniques mentioned is to manufacture items of the prescribed shape and properties.

CHAPTER 23

General

23.1. Plastic Deformation of Metals

Any change of shape or dimensions of a body is called *deformation*. It may be caused by both internal and external forces. Elastic deformation should be distinguished from plastic deformation.

Elastic deformation disappears as soon as the load is removed, while *plastic deformation* remains after the load has been removed.

Distinction should be made between hot and cold deformation of metals. These are, of course, conventional concepts, since it is impossible to state precisely where the boundary between cold and hot deformation lies. It is customary to apply the term *cold working of metals* to a process that involves no recrystallization of the metal (see Para 2.3). In *hot-working processes*, recrystallization occurs both during the deformation and immediately after

it, while the metal is still hot. There are numerous intermediate stages between these two extreme cases.

Subjected to initial plastic working are usually cast
blanks. Cast metal contains more or less pronounced blow
holes, zones of porosity, shrinkage cavities, etc., and its
plastic working compacts the material. Further working no
longer affects the compactness of the metal. This is one
of the principles underlying plastic working of metals, the
law of constant volume: the volume of metal after working is equal to that before working.

23.2. Quantification of the Deformations of a Body

One of the principal characteristics of deformation is
the *lengthening coefficient*, which is the ratio of the final
length of a body to its initial length.

When a body suffers a number of successive deformations,
its total deformation can be described in terms of total lengthening, i.e., the ratio of its final length to its initial length.

Linear lengthening is the absolute elongation of the body,
i.e., the difference between its length after working to
that before working.

Deformation of the body in height can be described in
terms of the *height increment*, i.e., the difference between
the body's heights before and after working, the ratio of
the heights, or the ratio of height increment to the full
height before or after the working. The change in the height
of the body is most conveniently expressed by natural
logarithm of the ratio of the height before working to that
after working.

Transverse deformation (deformation in width) is described by the value of *width increment*, i.e., the difference between the width of the body after working and before it,
or by the ratio of these widths.

Easiest to quantify is the deformation of a body of a
simple shape which retains its geometry after working.
This can be exemplified by bodies of square, rectangular,
round, or annular (tubular) cross sections.

In other cases, the degree of deformation is more complicated to estimate, and recourse is made then to the mean
value of complex deformation.

CHAPTER 24

Rolling

Rolling is one of the principal methods of plastic working of metals, employed to manufacture a great amount of square, round, and other simple and compound sections, rails, beams, sheet, pipes, etc. The cross section of a rolled item is termed *section* for short.

Rolling consists essentially in passing the metal between revolving rolls. Biting of the metal and further rolling rely on friction between the rolled strip and the rolls.

Fig. 57 illustrates schematically the process of longitudinal rolling. The *zone of deformation* is cross-hatched on the diagram. Angle α corresponding to this zone is termed the *angle of nip*. Arc *AB* is the *arc of nip*, whereas *l* is its projection upon the horizontal axis.

Rolling is a continuous and regular process, characteristic of which is an uninterrupted motion of the worked metal through the zone of deformation (between the rolls). Continuous processes yield products of uniform structure and properties, as the conditions of working of the metal are the same throughout the length of the product.

Continuous processes are very efficient, they can be conducted without stoppages. The breaks in operation, which occur practically, are due to either poor quality of equipment or poor operating procedures.

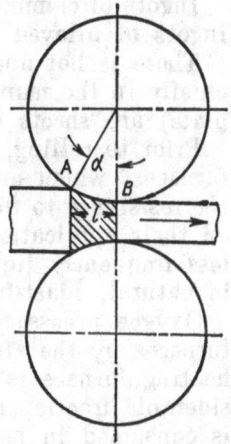

Fig. 57. Longitudinal rolling

The rolling process today consists of (1) manufacture of a semi-product from an ingot and (2) manufacture of a finished item from the semi-product. Each of these stages is subdivided into the following operations: preparation of starting materials, their heating, rolling proper, and finishing of the product.

24.1. Preparation of Ingots and Blanks for Rolling

The preparation of starting materials for rolling consists in the elimination of surface defects. This operation is particularly important in rolling of alloy steels. Surface defects are removed by means of swing grinders, air chippers, and torch scarfers. The former two methods are laborious and inefficient. Scarfing machines installed immediately downstream of the rolling mill are now used on a wide scale. These machines are capable of removing a layer of metal 3-6 mm thick from all the faces of the semi-product.

Also employed for cleaning blanks are milling machines which remove the entire surface layer of metal in the process of rolling proper.

Ingots of common carbon steels are not cleaned, whereas ingots of alloyed steels are cleaned in special machines.

There is hot and cold rolling. Cold rolling is employed chiefly in the manufacture of thin sheets; the blanks (strip plate) are sheets or strips obtained in hot-rolling mills.

Prior to rolling, ingots and blanks are heated in heating furnaces, which ensures their high plasticity and reduces the resistance to flow. The design of these furnaces depends on their application. The metal is heated usually by gas, less frequently liquid or solid fuel is used. The gas fuel is natural, blast-furnace, or coke-oven gas.

Oxygen necessary for fuel combustion is supplied in the furnaces by the air fed simultaneously with the fuel. If a heating furnace is supplied with cold air and gas, a considerable fraction of the heat generated in the combustion is consumed in heating them. To avoid these losses, the air and the gas (sometimes, only the air) are preheated in special arrangements termed *recuperators* and *regenerators*.

Large-size ingots are heated in *soaking pits* located in a bay of their own, adjoining the building which houses the rolling mill. The ingots are usually placed into the pits while still hot, this greatly increasing the throughput of the pits. Depending on the grade of steel, the ingots are heated to 1,120-1,300 °C. All operations in the soaking-pit bay are performed mechanically.

The current practice is to heat the semi-products (blanks) in *continuous billet-heating furnaces*. In such a furnace a pusher causes the metal to travel from one end of the furnace to the other. Hot gases moving in countercurrent above and below the metal gradually heat the ingots.

Cranes convey billets to the charging devices and the pusher feeds them into the furnace, advancing, in the process, other blanks which lie in the furnace in direct contact with one another. The billets are heated from top and bottom as they slide along skid pipes. These pipes are cooled by running water. Fig. 58 shows a schematic diagram

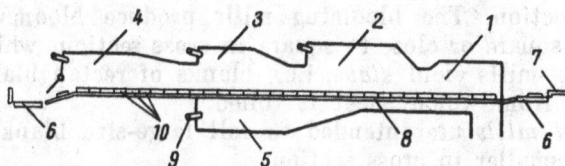

Fig. 58. Continuous billet-heating furnace

1 — preheating zone; *2,3* — heating zones; *4* — soaking zone; *5* — bottom heating zone; *6* — roll tables; *7* — slab pusher; *8* — skid pipes; *9* — burners; *10* — billets

of a four-zone heating furnace charged from one end. The temperature of ingots is raised in the preheating and heating zones, while the *soaking zone* is destined to equalize the temperature throughout the cross section of the billets. The billets are discharged through an end door.

Continuous furnaces may have two, three, four, or five zones. They may also be either end- or side-charged and discharged.

Billets are heated to 1,100-1,220°C.

Blanks may also be heated in *movable-hearth furnaces* (for large-size sheet ingots), *rotary-hearth furnaces* (blanks for pipe rolling), *conveyor-type furnaces* (strip blanks for sheet bars), etc. Some metal is always oxidized to scale (iron loss) during heating. In an efficiently operated furnace the iron loss is within 2 per cent.

Loss of metal in heating may be reduced by practicing low-scale or scale-free heating.

24.2. Rolling Mills

Classification of rolling mills. *A rolling mill* is a complex of machines and units directly intended for rolling, straightening, cutting, and conveying metal.

Rolling mills are classified according to application, arrangement of working stands, arrangement and number of rolls in the working stands.

By application, rolling mills fall into two large groups: cogging mills and finisher mills.

Cogging mills. *Blooming* and *slabbing mills* are large rolling mills intended to reduce ingots to blanks of a large cross section. The blooming mills produce blooms, i.e., blanks square or close to square in cross section, while the slabbing mills yield *slabs*, i.e., blanks of rectangular cross section from which sheet is rolled.

Billet mills are intended to roll large-size blanks into billets smaller in cross section.

Finisher mills. *Sheet mills* are intended to produce sheet steel from hot or cold blanks. They are characterized by the length of the roll body.

Rail-and-structural steel mills (or simply *structural mills*) are designed to roll rails, beams, channels, and other large sections.

Section mills are generally subdivided into heavy-section, medium-section, and small-section varieties. The principal characteristic of a section mill is the diameter of rolls in the last stand. Section mills produce rounds, squares, angles, beams, channels, die-rolled sections, etc.

Wire mills serve to roll wire and rod from 6 to 10 mm in diameter.

Pipe and tube mills are employed to produce pipes and tubes.

Arrangement of working stands. By the arrangement of working stands, rolling mills may be divided into single- and multi-stand mills.

Single-stand mills are very widely used. Here belong blooming, slabbing, sheet, and other rolling mills. A schematic diagram of a single-stand mill is shown in Fig. 59a.

Multi-stand mills may be arranged either side by side or in tandem.

Mills with the side-by-side arrangement of the stands (Fig. 59*b*) are commonly used as rail-and-structural steel and heavy-section mills.

The side-by-side mills are less costly, but have a substantial drawback. Roll speed is the same in all the stands; as strip length increases after each pass, the final stand becomes a bottleneck. Because of this, the rolling rate in these mills is quite low.

Rolling speeds in subsequent stands may be increased by arranging the work stands in two, three and sometimes even four lines (Fig. 59*c*). This arrangement is used in rail-and-structural steel, wire, and small-section mills.

Fig. 59*d* shows a two-stand mill with stands arranged in tandem. This kind of mills is used to roll plates, the piece being passed a number of times through each stand.

Continuous mills (Fig. 59*e*) are, the most advanced type. Their stands are arranged in tandem and in each stand the metal is rolled only once. Rolling speed increases in subsequent stands, this being obtained by proper selection of roll diameters and gear ratios (for group-driven stands). There are also mills with an individual drive for each stand. The spaces between the stands are considerably less than the length of strip, therefore, the strip is worked in several or even in all of the stands at a time. These are very efficient mills. They are used to roll blanks, sheets, bands, and small sections.

The rolled metal may be stretched or squeezed between the stands, this being a substantial shortcoming of the continuous mills. Squeezing and stretching cause a change in the dimensions of the strip.

A zigzag arrangement of the stands (Fig. 59*f*) eliminates this shortcoming; here, the strip is passed only once through each stand, but it is in only one stand at a time. As is seen from Fig. 59*f*, the stands are arranged in three lines, which prevents the plant building from being excessively long. All the three strands are located in a single bay.

Also used are mills with staggered arrangement of the finishing stands. A schematic diagram of such a mill is given in Fig. 59*g*.

According to the arrangement and number of the rolls, stands are classified as follows.

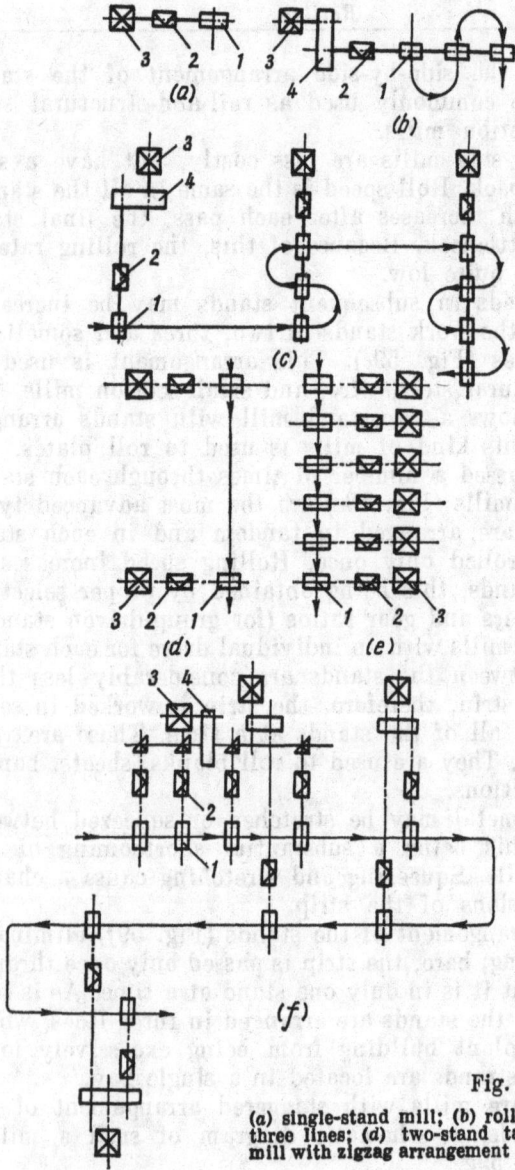

Fig. 59. Rolling

(a) single-stand mill; (b) rolling mill with three lines; (d) two-stand tandem rolling mill with zigzag arrangement of stands; (g) ectric motor:

Two-high stands—the most widely used ones at present —have two work rolls. They may be either reversible or non-reversible. The rolls of non-reversible stands rotate at a constant speed in one direction. These stands are used in continuous, tandem, and staggered-stand mills. The rolls of reversible two-high stands rotate alternatively in opposite directions at varying speeds. These stands are used in blooming, slabbing, and plate mills. The axes of both rolls of two-high stands are horizontal (Fig. 60*a*).

Double two-high stands are equipped with two pairs of horizontal rolls. The arrangement of the roll pairs is shown in Fig. 60*b*. The strip is rolled alternatively in each pair of the rolls. This is an outdated design, and no newly-built stands have such an arrangement of the rolls.

Three-high stands have three horizontal rolls, revolving always in the same direction (Fig. 60*c*.) The strip is passed forward between the top and middle rolls, then reversed between the middle and bottom rolls. The strip is lifted

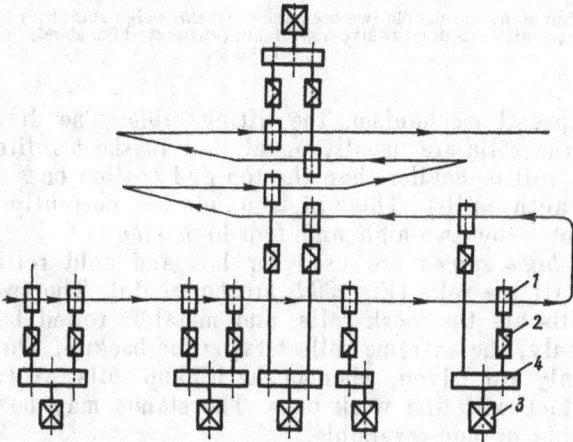

(g)

mill arrangements
side-by-side arrangement of stands; (c) rolling mill with stands arranged in mill; (*e*) continuous rolling mill with individually driven stands; (*f*) rolling rolling mill with staggered stands; *1* — working stand; *2* — pinion stand; *3* — el- *4* — gearbox

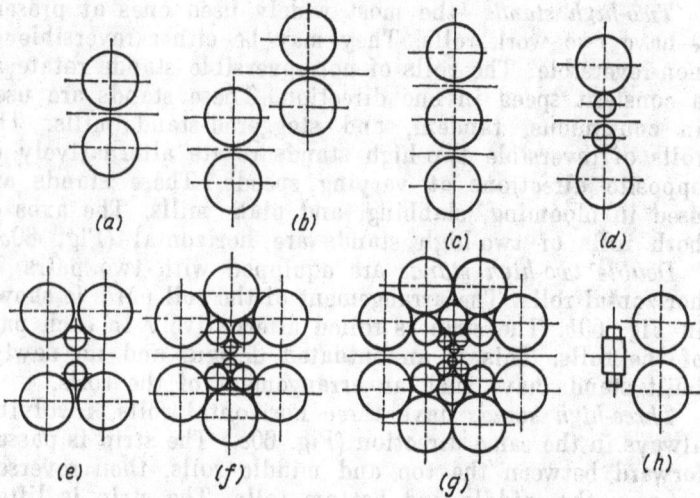

Fig. 60. Arrangement of rolls in work stands of rolling mills

(a) two-high stand; (b) double two-high stand; (c) three-high stand; (d) four-high stand; (e) six-high stand; (f) twelve-high stand; (g) twenty-high stand; (h) universal stand

by a special mechanism, the lifting table. The diameters of all the rolls are usually equal, but in sheet rolling the middle roll is smaller than the top and bottom ones (three-high Lauth mills). Three-high stands are currently being supplanted by two-high and four-high stands.

Four-high stands are used for hot and cold rolling of sheet. All the rolls (Fig. 60d) are horizontal. The two central rolls are the work rolls, and metal is rolled between them only, the extreme rolls serving for backup. The work rolls only are driven, whereas the backup rolls revolve due to contact with the work ones. The stands may be either reversible or non-reversible.

Multiroll stands may be of six-, twelve- and twenty-high design (Fig. 60e through g). Each of these stands has only two work rolls, the others being backup rolls. In recent years, twelve- and twenty high stands have gained wide recognition. They are highly rigid and ensure the rolling of very thin band.

Universal stands have two pairs of rolls, one horizontal, and the other vertical (Fig. 60*h*). The vertical rolls are located before and after horizontal ones, and they may be either live or idle. The universal stands are employed for rolling slabs, sheet, beams.

Vertical stands have two vertical rolls. They are used in continuous mills and are arranged alternatively with the horizontal stands to avoid the tilting of the strip.

Stands of special types employed for rolling pipes, wheels, and wheel tyres have various numbers of rolls.

General arrangement of a rolling mill. The roll line of a rolling mill consists of working stands *1*, spindles *7*, pinion stand *2*, gearbox *3*, flywheels *5*, couplings *6* and electric motor *4* (Fig. 61). Commonly, some of the above

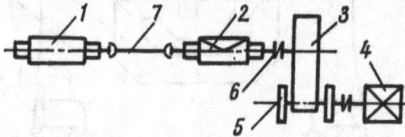

Fig. 61. Mill roll line

1 — working stand; *2* — pinion stand; *3* — gearbox; *4* — electric motor; *5* — flywheel; *6* — coupling; *7* — spindle

elements are omitted. Thus, a blooming mill with individually driven rolls has no pinion stand, no gearbox, and no flywheel. In group-driven continuous mills, one electric motor drives a number of stands via a single gearbox.

Fig. 62 shows a schematic diagram of a work stand of a two-high continuous billet mill. The shape of work rolls depends on the application of the rolling mill. Fig. 63*a* shows a sheet roll, Fig. 63*b*, a blooming-mill roll. Rolls are made from steel or cast iron. They are replaced at regular intervals, as they wear out owing to contact with the rolled metal. Worn rolls are reconditioned by turning on roll lathes and re-installed in the mill. Built-up welding of rolls has gained a wide recognition in recent years. As the rolls are machined, their diameter decreases, and there comes a moment when they can fail. In case of built-up welding, the diameter of the rolls remains unchanged,

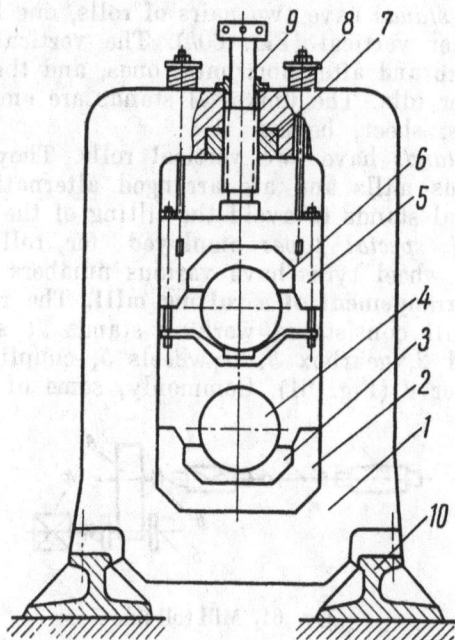

Fig. 62. Two-high working stand

1 — housing; 2 — bottom chock; 3 — bottom bearing; 4 — work roll; 5 — top chock; 6 — top bearing; 7 — nut of screwdown arrangement; 8 — spring; 9 — housing screw; 10 — mill shoe

since the roll passes are each time restored by welding. This increases substantially the roll service life.

The rolls are mounted in two housings which receive the pressure the metal exerts upon the rolls in the process of rolling. The housing (Fig. 62) is a frame-shaped rigid metal casting. The two housings of a rolling stand are rigidly secured at the top and at the bottom by special bolts and spacer pipes or by cast housing separators. The feet of the housings are set on mill shoes and bolted to them. Mill shoes are long bars of an intricate configuration, usually cast from iron and, sometimes, from steel.

Necks of mill rolls (Fig. 63) revolve in bearings or in bushings. Split or pressed solid bushings are most frequently made of textolite. Textolite features high durability (in case of a correct service), requires no special lubricants as it is lubricated by water, and has a low coefficient of friction.

In recent years, rolling-contact and film-lubrication bearings have received extensive application. The rolling-con-

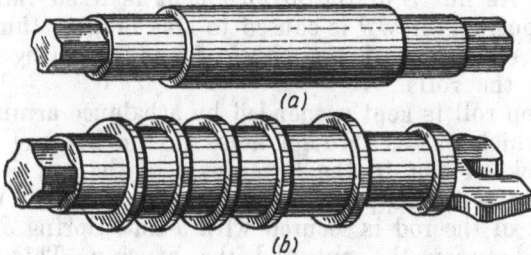

(a)

(b)

Fig. 63. Sheet (*a*) and blooming (*b*) mill rolls

tact bearings commonly used are of the double-row cone type.

The bushings and the bearings are mounted in mill chocks which rest directly on the housing. The chocks are fastened by bolts and plates to the housing and lock the rolls securely in a horizontal plane during rolling. The top roll is adjustable vertically by means of screwdown and balance arrangements. The bottom roll does not commonly move in the vertical direction.

The top roll is adjusted either after each pass or at given intervals, depending on the rolling procedures and the mill application.

In reversible mills (blooming, slabbing, sheet, and other mills), the same workpiece is rolled several times in a single stand, becoming thinner and thinner as the rolls are brought closer together after each pass. In non-reversible mills, where the strip passes each stand only once, the roll gap must be adjusted either in case of a change in the thickness of the product to be rolled (as, rolling of a 2-mm sheet after the production of a 10-mm sheet) or in case of wear of the roll surface, this requiring the drawing of the rolls together. In fact, the movement of the top roll is necessary for adjustment of the stand. The top roll is caused to travel by means of the screwdown and balance arrangements.

The screwdown and balance arrangements are greatly varied in construction. Fig. 62 illustrates such a device of an elementary design.

The screwdown arrangement may be a hand-operated one, in which the screw is rotated manually with a spanner or a lever. As nut 7 of the arrangement is fixed immovably in the housing, screw 9 is caused to sink or rise, thus changing the elevation of top chock 5 and adjusting the gap between the rolls.

The top roll is kept suspended by a balance arrangement, a rod which passes through holes in the housing and the top chock. A slot in the lower part of the rod receives a cotter pin which supports the top chock, while the threaded top part of the rod is secured with a nut. Spring 8 is compressed between the nut and the housing. This type of gear is called the spring balance arrangement.

The above design of screwdown and balance arrangements is employed in mills which do not require frequent adjustments of the gap between the rolls. Modern large-output rolling mills are fitted with electrically-driven screwdown gears which ensure a rapid adjustment of the roll.

In recent years, most generally employed in mills are hydraulic balance arrangements described below. A cylindrical chamber in the bottom chock receives a piston from above. High-pressure oil fed into the chamber raises the piston until it thrusts against the top chock. The top chock is thus locked between the piston and the housing screw.

In those housings where the top roll is adjusted after each pass, only electric screwdown gears are employed. The same electric motor balances the top roll by means of a reverse screw on a worm wheel. The reverse screw has a left-handed thread of the same pitch as that of the housing screw, and because of this the chock repeats exactly the motion of the housing screw. The housing screws are made from steel, the nuts are from bronze.

To direct the rolled metal into and out of the rolls, the rolling mills are equipped with special leading bars and troughs termed *guides*. There are entry and exit guides. Their shape is suited to the cross section of the rolled me-

tal. The exit guides have one more purpose. Metal has to
be tilted during rolling, and this is done by *helical* (with
a helical internal surface) or *roller* (the tilting action provided by rollers) *twister guides.*

The next major element of the rolling mill roll line is
the pinion stand. It is omitted only in mills with indivi-
dually-driven rolls and in mills where only one roll is driven.

The destination of the pinion stand is to transmit rotation from one shaft to all the work rolls: two in two-high,
three in three-high stands. As the rolls must rotate at the
same speed, the gear ratio of the main gears is always equal
to unity. The gears are of the herringbone type.

The main *gearboxes* of the mill serve to step down the
speed of the rolls when a high-speed electric motor is employed (this type of motor is less costly). Commonly, single-
stage gearboxes (one pair of gears) are used in rolling mills;
two- and three-stage (respectively, two and three pairs of
gears) gearboxes are also employed sometimes.

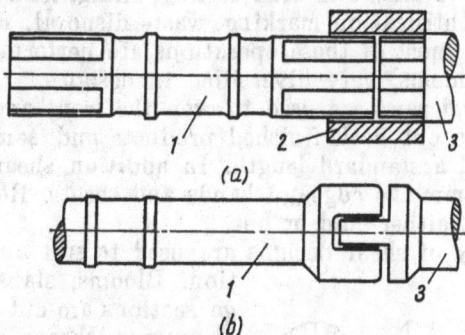

Fig. 64. Rolling-mill spindles

(a) wobbler type; (b) universal type
1 — spindle; *2* — coupling; *3* — work roll

Gaining recognition are *combination pinion stands* in
which the pinion stand and the gearbox are accommodated
in a single housing. The *flywheels* mounted on the shaft
of the gearbox pinion serve to accumulate energy in intervals between the rolling of strip. If these intervals are
short, there is no point in using the flywheels.

Gears of the pinion stand are connected to the work rolls by coupling spindles (Fig. 64). In side-by-side mills, the coupling spindles interconnect the work rolls of adjacent stands. The spindle is coupled to the roll wobbler by a coupling. There is a wide variety of spindle designs. In mills with a considerable travel of the top roll (such as blooming, slabbing, and plate mills) universal spindles are used (Fig. 64b).

Couplings of various designs are employed to interconnect the shaft of the electric motor and the driving shaft, the driving shaft and the gear of the pinion stand, the spindles and the work rolls.

24.3. Rolling Mill Auxiliaries

The operation of a rolling mill involves, besides rolling, a host of auxiliary operations, including transportation of stock to the rolling mill, handling of strips between the rolling-mill stands and after rolling, straightening, cutting, reeling, identification marking, waste disposal, etc. In modern mills, most of these operations are performed by special mechanisms, very diversified in design.

Shears and saws are used to crop the front and the back ends and to cut both finished products and semi-products to pieces of a standard length. In addition, shears are employed to trim the edges of bands and sheets. Rolled stock may be cut either cold or hot.

A variety of shear designs are used to suit any application. Blooms, slabs, and large sections are cut by heavy-duty crop shears capable of shearing squares 450×450 mm and higher, but only when these are hot. The blade arrangement in such shears is diagramed in Fig. 65a.

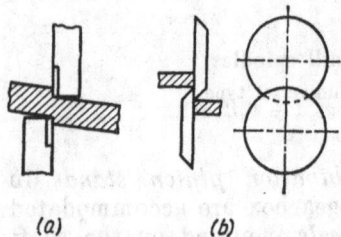

Shears with a similar arrangement of blades, but considerably smaller in size, are used for cold cutting of small sections. Strip, sheet, and

(a)   (b)

Fig. 65. Arrangement of blades in crop shears (a) and in circular shears (b) small sections are cut in packs by similar shears, but the cutting edges of the top and bottom blades are set at an angle.

Fig. 65*b* illustrates schematically the arrangement of blades in circular shears which are used to trim the longitudinal edges of sheet and band and to slit these products into strips.

Continuous mills yield products of a very great length, which must be cut without stopping the mill. The so-called *flying shears* meet this requirement.

When cut by shears, the edges of rolled stock are somewhat crumpled, and the plane of the cut may be ragged. If a straight cut and uncrumpled edges are required, use is made of *saws*. The cutting tool of a saw is a toothed disk rotating at a great speed.

Reels are used to wind the rolled metal into coils. Usually this is required for strips

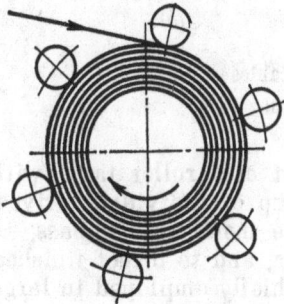

Fig. 66. Winder reel for strip

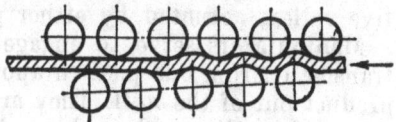

Fig. 67. Straightening

and rolled stock of a small cross section. Fig. 66 shows a schematic diagram of a reel for a wide hot-rolled strip.

Straightening machines. Rolled products are levelled chiefly in roller-straightening machines, the process being as shown in Fig. 67. This method is used for flattening sections and sheet metals.

Roll tables are machines for longitudinal transportation of metal. They consist of a number of rotating rollers powered by a group or individual drive. Roller tables are divided into main and carry-over varieties. The main tables are located at the rolling stand.

Lifting tables are employed chiefly in three-high mills for feeding the strip between the central and top rolls. There are *parallel and tilting tables* (Fig. 68). In a tilting-type table only the side nearest to the rolls rises. They are widely used since they are capable of handling strips of any length.

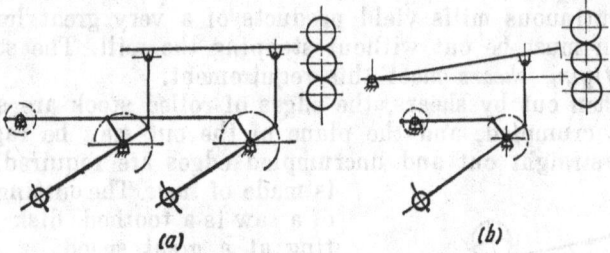

Fig. 68. Lifting tables

(a) parallel type; (b) tilting type

The lifting tables usually consist of a roller table with live rollers actuated by either group or individual drives.

Manipulators serve to engage the metal into a pass, to transfer it from one pass to another, and to direct finished product out of the mill. They are chiefly employed in large reversible mills, such as blooming and slabbing mills and breakdown stands.

A manipulator is essentially composed of two guides arranged in parallel to one another.

Tilters are used to tilt rolled strip. They are either several hooks (three or four in number) set on a single shaft, or a tilting bushing. Tilters of the former type are employed in blooming and slabbing mills and breakdown stands; they are commonly arranged on the manipulator guides. Tilters of the latter type are used in zigzag mills.

Cooling banks are intended for cooling rolled metal. While cooling, the metal moves at a certain speed transversally with respect to the direction of rolling. Cooling banks with two systems of racks are the most widely used ones. Coiled metal is handled by hook conveyors which serve as cooling facilities.

Rolled stock preparation and finishing lines consisting of a host of various mechanisms are now being extensively employed.

24.4. Range of Rolled Products

The totality of sections and their sizes, rolled in a given mill or a group of mills, is termed the *range of rolled products*. Rolled stock may be divided into four principal groups: sheet steel; section steel; pipes and tubes; and special kinds of rolled stock.

Sheet steel is described by its dimensions: thickness and width. In recent years, the output of sheet steel has been on the rise. Section steel is greatly varied in shape and size. Fig. 69 shows some of the steel sections. Pipes and tubes are produced in a variety of cross sections and sizes (Fig. 70). They may be either seamless or welded.

Special kinds of rolled products include die-rolled sections, tyres, wheels, balls, etc.

Formed sections manufactured from sheet and band steel in special machines have gained a wide recognition. They have a number of advantages as compared to rolled sections. First, formed sections may be obtained in shapes which cannot be manufactured by hot rolling. Second, formed sections are more accurate in dimensions and have a smaller thickness.

24.5. Fundamentals of Roll Pass Design

The surface of a roll is provided with a recess termed *groove* of a shape corresponding to the cross section of the product to be rolled. The grooves of a pair of rolls form a *pass*.

The term *roll pass design* denotes the sequence of reductions in each pass and the series of roll passes necessary to manufacture the section in question.

The overall deformation is usually found from the required dimensions of the finished section and the known size of the starting blank or ingot. Therefore, the problem lies in determining the number of passes and the reduction to be obtained in each pass.

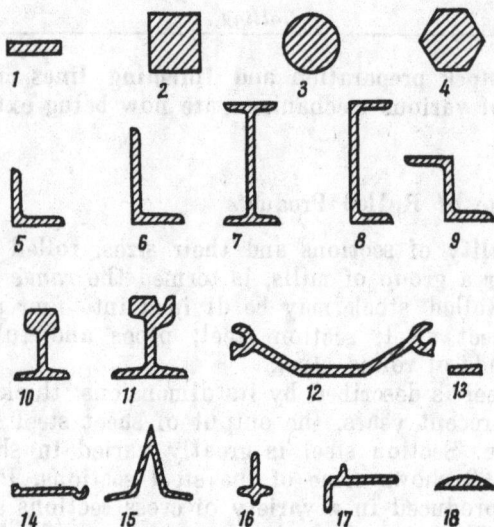

Fig. 69. Steel sections

1 — strip; *2* — square; *3* — round; *4* — hexagon; *5* — equal angle; *6* — unequal angle; *7* — I-beam; *8* — channel; *9* — Z-section; *10* — railway rail; *11* — tramway rail; *12* — piling bar; *13* — wedge; *14* — automotive rim; *15* — tractor track; *16,17* — window-sash sections; *18* — segment

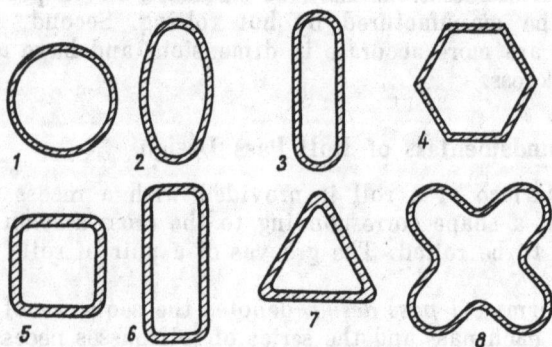

Fig. 70. Pipe and tube sections

1 — round; *2* — oval; *3* — flat-oval; *4* — hexagon; *5* — square; *6* — rectangular; *7* — triangular; *8* — club-shaped

The pass reduction depends on the following factors: biting of the metal by the rolls, strength of the rolls, power rating of the electric motors, roll wear, and plastic properties of the metal.

Roll pass design for sheet rolling. In order to obtain a sheet of an even transverse thickness, the roll gap must be uniform throughout the length of the roll body. In sheet rolling the pressure of the metal upon the rolls is so great that the rolls bend, the rolls gap taking the shape of a lens. In addition, the rolls are heated due to contact with a hot band in hot rolling and to the heat generated in cold-rolling deformation, this leading to thermal expansion of the rolls and distortion of the gap. Both of these phenomena affect the gap shape, each in its specific manner, and they should be given due consideration.

If roll gap is lens-shaped (convex or concave), the cross section of the finished sheet will be shaped accordingly. To avoid this, the rolls are given a slightly convex or concave shape. In cold-rolling mills where the thermal convexity is small, the rolls are made convex to compensate for the bending. In hot-rolling mills, the rolls are concave as the thermal deformation exceeds the bending.

The starting convexity or concavity (*roll contouring*) of rolls is determined experimentally and by calculations.

Practice shows that rolling is steadier when the band is slightly cambered. Therefore, the rolls are so contoured as to produce a slight convexity in the cross section of rolled band.

Roll pass design for rolling blanks, section steel, and structural steel. Blanks, section steel, and structural steel are rolled in a number of successive passes in which the strip is drawn and its cross section reduced to obtain the required finished section.

By their function, all passes are divided into reducing (breakdown), preleader (former), and finishing (final) types. *Breakdown passes* are variously shaped (Fig. 71). Breakdown passes are intended for a high-rate drawing of the strip.

In the *preleader passes*, the strip is drawn down still further and is gradually formed to the shape of the finished product. In the *finishing passes*, the ready section is given its final shape.

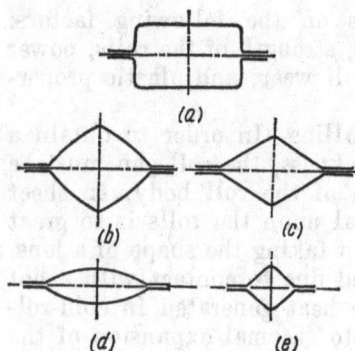

Fig. 71. Breakdown passes

(a) box-type; (b) gothic; (c) diamond;
(d) oval; (e) square

Let us give a detailed consideration to the roll pass design in various rolling mills. Blanks are rolled in blooming or billet mills. A blooming-mill roll has one large and straight slightly recessed groove (the so-called roll *barrel*) and a number of box passes (Fig. 72).

The advantages of the box passes are as follows:

(1) the stock can be passed a number of times through each of them, the rolls being brought closer together after each pass;

(2) the depth of the groove in the roll body is small;

(3) the groove affects the roll strength but slightly.

The grooves may also be arranged along the length of the roll in a different manner, the smooth portion (barrel) being in the middle and the box passes on both sides of it.

Ingots are the starting material for a blooming mill. Initially, the material is rolled on the barrel, then successively in each of the passes. The workpiece is tilted at

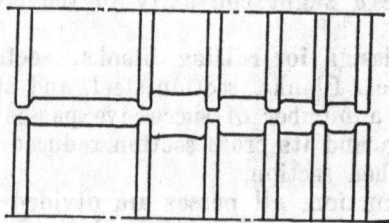

Fig. 72. Grooves in blooming mill rolls

regular intervals, after each even-numbered pass; an infrequent practice is to tilt workpieces at greater intervals (after a fourth or sixth pass) at the early stages of rolling, and to perform tilting after each other pass later.

Blanks are rolled chiefly in continuous billet mills composed of several stands. Box passes are employed in the starting stands, followed by the diamond-diamond and the diamond-square series.

Fig. 73 illustrates the passes of a continuous billet mill using the diamond-square series. The clearance between the collars of the top and bottom rolls allows the size of the passes to be adjusted within certain limits.

The four rows of passes produce blanks in three sizes without a roll change-over. Blanks sized 55×55 mm are the most often rolled ones, so two rows of passes are provided for them.

Blooms and blanks are rolled into plain shapes (round, square, strip) and sections (rails, beams, angles, channels, etc.).

In mills intended for the rolling of plain or light sections, the first few stands serve to draw the strip further; used in these stands are the above mentioned series of breakdown passes or the oval-square series, where the square passes alternate with the oval ones. This series produces the most intensive breakdown, it is most often used in the preleader groups of light-section and wire mills.

Rounds are finally rolled by the square-oval-round or round-oval-round method, while the squares are rolled by the square-diamond-square and sometimes square-diamond-diamond-square methods. The strip is tilted through 45 or 90° after each pass.

Wear of preleader-pass rolls has little effect upon the shape of the final product, whereas wear of the prefinishing or finishing passes may result in a substantial deviation of the section from the required shape and thus give rise to rejects. Therefore, the rolls of the finishing-group stands are provided with a number of identical passes. As one of the passes wears out, the stock is rolled in another one. Breakdown passes of the preleader stands may be used for the rolling of a number of shapes.

Steel strip in widths up to 200 mm may be rolled in passes, in step rolls or in universal mills. At present, building of special strip rolling mills is considered unpractical. To obtain strip, a wider strip rolled in continuous sheet mills is slit (cut along) in rotary gang slitters.

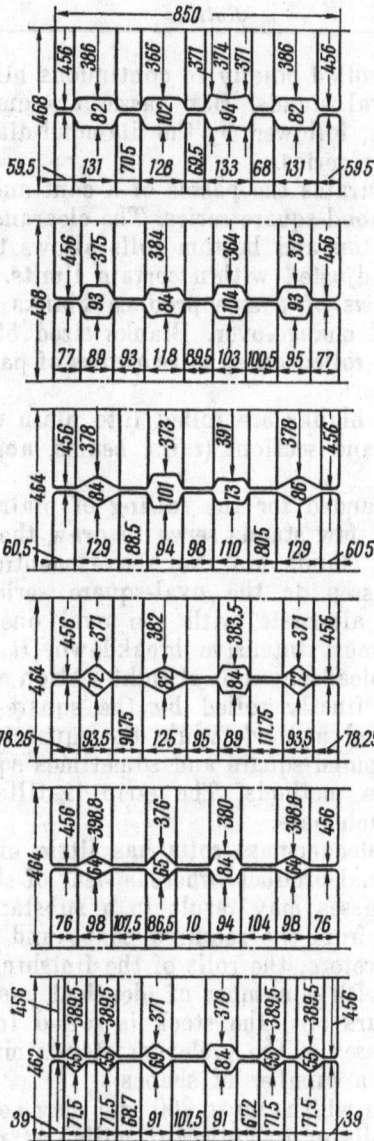

Fig. 73. Roll pass design in continuous billet mill

Characteristic of the rolling of sections (angles, beams, channels, etc.) is an unequal deformation of different parts of the sections. This non-equality is due to the fact that each of the sections mentioned is rolled from a rectangular blank.

In the initial passes, where strip temperature is still high and the metal is more plastic, the strip is given a shape close to that of the finished section, then it is gradually finished in the other passes. Fig. 74 illustrates the pass design for a $100\times100\times6.5$ angle.

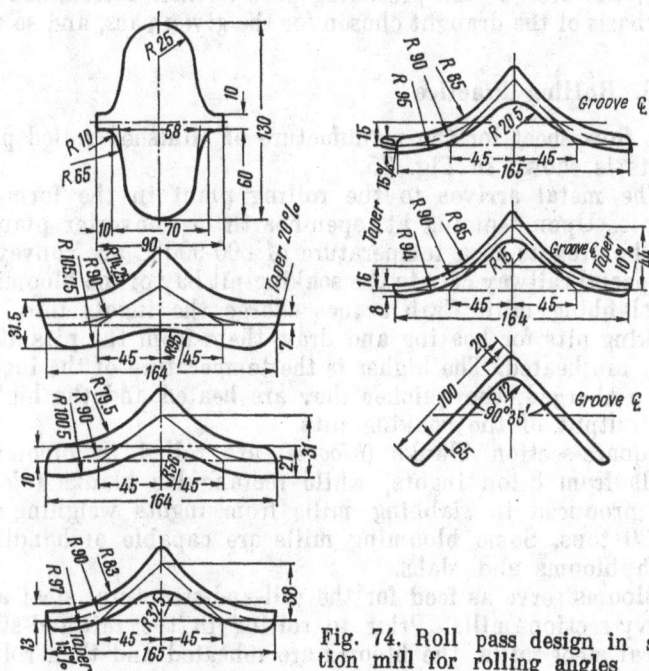

Fig. 74. Roll pass design in section mill for rolling angles

The roll pass design for any type of section requires the determination of the following values:

(1) Overall draught, calculated on the basis of the sizes of the finished section and those of the blank (sometimes

the sizes of the finished shape only are known, while the size of the blank must be determined).

(2) Required number of passes, taken as factor of the mean deformation which, in turn, depends on the roll biting capacity and other practical data.

(3) Draught in each pass, determined chiefly on the basis of practical data.

(4) Pass dimensions.

Generally, the procedure of pass design starts with determining the dimensions of the finishing pass with due regard for the cooling shrinkage of the metal after the rolling; the area of the preceding pass is then determined on the basis of the draught chosen for the given pass, and so on.

24.6. Rolling Practice

A flow sheet for the manufacture of finished rolled products is shown in Fig. 75.

The metal arrives to the rolling plant in the form of ingots. Upon teeming at open-hearth or converter plants, the hot ingots at a temperature of 600-900°C are conveyed in special railway cars to the soaking-pit bay of the blooming or slabbing mill. Crab cranes charge the ingots into the soaking pits for heating and draw them from the pits after they are heated. The higher is the temperature of the ingots when charged, the quicker they are heated and the higher the output of the soaking pits.

Square-section blanks (*blooms*) are rolled in blooming mills from 8-ton ingots, while rectangular blanks (*slabs*) are produced in slabbing mills from ingots weighing up to 30 tons. Some blooming mills are capable of handling both blooms and slabs.

Blooms serve as feed for the rail-and-structural steel and heavy-section mills. Prior to rolling in the rail-and-structural steel mills, the blooms are reheated and then rolled first in a two-high reversible breakdown stand with rolls 900-950 mm in diameter, then in three stands arranged side by side; usually two of these are of a three-high type with a common drive, while the final two-high finishing stand has a separate drive. The stands may also be arranged in a different fashion. Rolled rails and beams are cut to

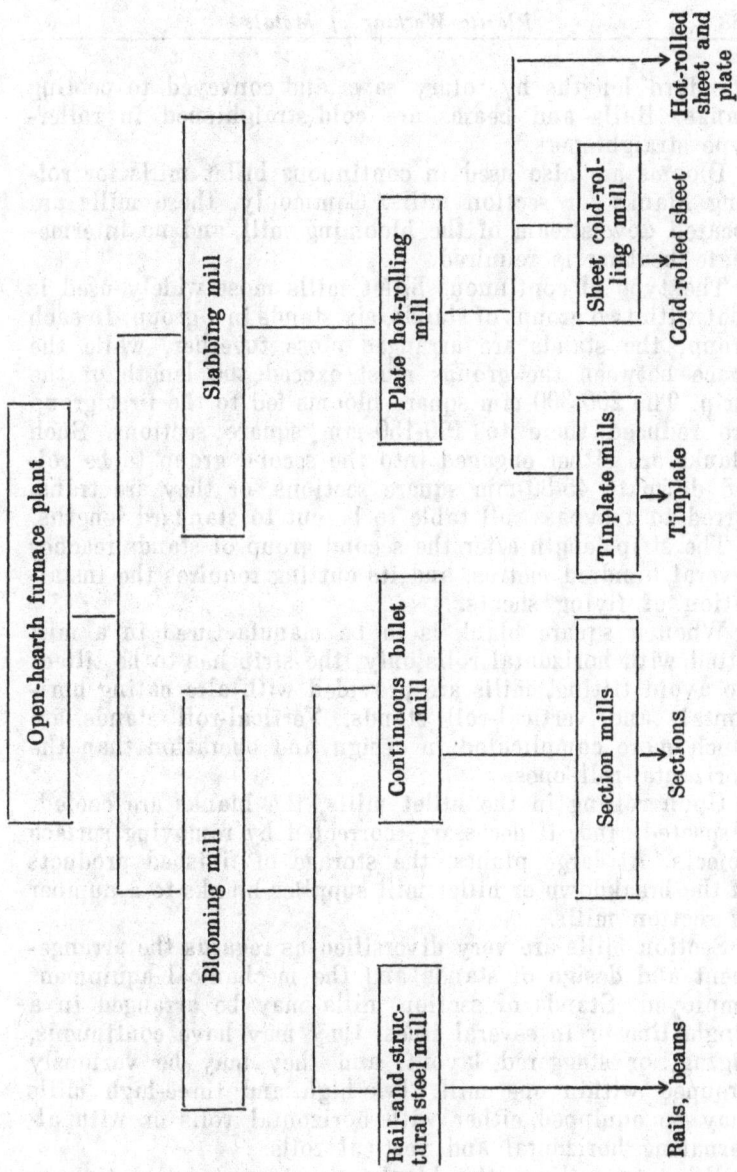

Fig. 75. Flow sheet of rolled stock manufacture

standard lengths by rotary saws and conveyed to cooling banks. Rails and beams are cold-straightened in roller-type straighteners.

Blooms are also used in continuous billet mills for rolling blanks for section mills. Commonly, these mills are located downstream of the blooming mill, and no intermediate heating is required.

The type of continuous billet mills most widely used is that with two groups of stands, six stands in a group. In each group, the stands are arranged close together, while the space between the groups must exceed the length of the strip. The 200-300-mm square blooms fed to the first group are reduced there to 100-150-mm square sections. Such blanks are either engaged into the second group to be rolled down to 45-80-mm square sections, or they are transferred to a bypass roll table to be cut to standard lengths.

The strip length after the second group of stands reaches several hundred metres, and its cutting requires the installation of flying shears.

When a square blank is to be manufactured in a mill fitted with horizontal rolls only, the strip has to be tilted. To avoid tilting, mills are provided with alternating horizontal- and vertical-roll stands. Vertical-roll stands are much more complicated in design and operation than the horizontal-roll ones.

Upon rolling in the billet mills, the blanks are cooled, inspected, and, if necessary, corrected by removing surface defects. At large plants, the storage of finished products of the breakdown or billet mill supplies blanks to a number of section mills.

Section mills are very diversified as regards the arrangement and design of stands and the mechanical equipment employed. Stands of section mills may be arranged in a single line or in several lines, they may have continuous, zigzag, or staggered layout, and they may be variously grouped within one mill. Two-high and three-high mills may be equipped either with horizontal rolls or with alternating horizontal and vertical rolls.

Prior to rolling, the blanks are heated in continuous furnaces and delivered to the receiving roll table. Depending on the design and location of the stands, the stock is rolled in each stand either once only or several times, using different passes. The stock is tilted with the aid of various types of guides or tilting gears.

The temperature of the strip should be maintained within a prescribed range during each rolling operation. In blooming and continuous billet mills, a low temperature of rolled pieces may destroy the equipment and overload the electric motors; in rail-and-structural steel and section mills, too low a temperature may also affect the quality of the finished product.

Upon rolling, the sections are cooled in the cooling banks, straightened, and cut to standard lengths, then inspected and the defects corrected at the stocking yard.

In wire and in some of the light-section mills the product is reeled and transferred to a hook conveyor which also serves as the cooling bank. In this case the wire (or rod, as it is more often called) and the light sections are shipped to consumers in coils.

Slabs are the starting material for sheet rolling. The slabs are produced in blooming or slabbing mills. The slabbing mills differ from the blooming mills in that they have a pair of vertical rolls located either at the front or rear side of the mill. Each horizontal roll is driven by an individual electric motor.

A continuous sheet hot-rolling mill is erected downstream of a slabbing mill.

The hot-rolling mill for producing 2 to 11-mm sheet comprises a roughing and a finishing mill group. In the former there are four or five stands arranged in tandem. When the sheet must be wider than the slab, first comes a spreading stand where the slab is rolled transversely (spread in breadth). The other stands of the roughing group are of the universal type, i.e., they are provided additionally with a pair of vertical rolls. The stands of the roughing group are spaced in a manner to allow rolling of workpieces in one stand only at a time.

The finishing group consists of six or seven four-high stands located side by side, the workpiece being rolled simultaneously in all of the stands.

High quality of surface finish is essential to sheet products, so slabs are thoroughly descaled by two scale breakers, one upstream of the roughing group, the other upstream of the finishing group. The scale breakers are two-high stands which slightly cog the feed and thus break up the scale. During rolling scale is removed by water ejected from high-pressure hydraulic sprays.

Upon rolling, the strip is reeled and transferred either to the finished products storage or to the sheet cold-rolling or tinplate plant.

Before rolling, a hot-rolled strip is treated in a continuous strip pickler which removes scale.

Upon pickling, the strip is cold-rolled in three- or four-stand non-reversible continuous mills where the strip is rolled down to a thickness of 0.4-2.5 mm.

Tinplate 0.15-0.5 mm thick is rolled in five- or six-stand continuous mills, all the stands being of the four-high type.

Strip is cold-rolled at high speeds. Thus, the speed in the final stand of a four-stand mill may reach 20 metres per second, and that of a five-stand tinplate mill, 35 metres per second.

Continuous multi-stand mills are built at large plants. When a relatively low output is contemplated, sheet is cold-rolled in single-stand four-high reversible mills equipped with reels. They produce a strip as thin as 0.5 mm in a few passes. Twelve- and twenty-high mills are used to handle hard-to-roll steels.

Metal is work-hardened in cold rolling. To counter this adverse effect and to obtain the necessary mechanical properties in cold-rolled sheet, the coils are annealed in bell-type furnaces. Annealing is followed by temper rolling. This is a conventional rolling of the strip in a single-stand four-high mill, but with a small reduction. The temper rolling improves the surface finish and mechanical properties of the sheet. Upon temper rolling, the coil is transferred to finishing units where the edges are trimmed and the coil is cut to standard sheets shipped in packs to consumers.

Cold-rolled strip, tinplate included, may be given various protective coatings. Widely used at present are galvanizing and tinning; coating with aluminium, chromium, plastics and varnishes is being employed on an ever wider scale. Coatings are applied in special processing lines.

Let us examine plate rolling procedures. It should be noted that the terms sheet and plate are to a great extent conventional. It is generally assumed that anything thicker than 4 mm is a plate. Thus, some of the plate is rolled in continuous sheet mills.

Plates in thicknesses from 4 to 50 mm are rolled chiefly in two-high reversible mills. The blanks—slabs and, sometimes, ingots—are preheated in continuous furnaces before rolling.

The stands are arranged side by side, the rolls of each stand being provided with individual drives. The stands may be associated in number of ways, as: a two-high stand followed by a four-high one (a combination in general use now); a two-high and a three-high Lauth stands (in the latter, the middle roll is idle and of a smaller diameter than the top and bottom rolls); both stands are four-high; or both stands are three-high Lauth type. The first stand (the roughing one) may be preceded by a vertical-roll stand, the finishing stand being of the universal type. The strip is first rolled in the roughing stand, several passes being made with the rolls brought closer together after each pass. The workpiece is then rolled in the finishing stand where several passes are also made.

Upon rolling, the plate is worked in straighteners and cooled on roller tables and cooling banks. The front and back ends are cropped by shears, and the plate is also cut to standard lengths.

Rolling plants are provided with a heat treatment department (for annealing, tempering, normalizing of sheet and plate) and pickling department.

Plate may also be manufactured in single-stand mills of various designs. Plates thicker than 60 mm are rolled in two-high reversible mills with rolls up to 4,500 mm long.

24.7. Pipe Rolling

Manufacture of seamless pipes. The manufacture of seamless pipes includes the following operations:

(1) manufacture of shells from blanks (piercing);

(2) manufacture of pipes from shells (rolling);

(3) finishing of the pipes.

The starting material for pipe manufacture are round-section blanks. The blanks are heated in rotary-hearth furnaces or in continuous sloping-hearth furnaces.

Blanks are pierced in piercing stands. The working tools in these mills are rolls of various shapes (Fig. 76).

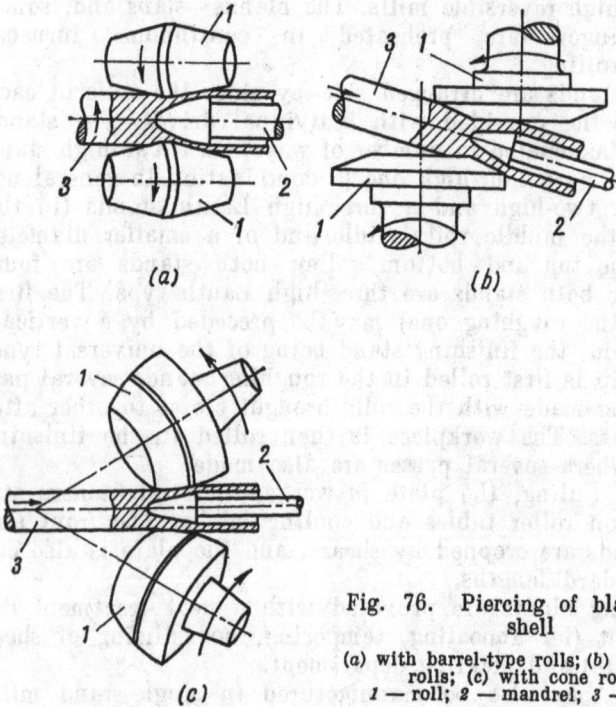

Fig. 76. Piercing of blank into shell

(a) with barrel-type rolls; (b) with disk rolls; (c) with cone rolls

1 — roll; 2 — mandrel; 3 — blank

The rolls are so shaped as to impart a rotary and translational motion to the blank causing so high a stress at its centreline that the metal there is torn apart and produces an axial hole. This hole is enlarged to the required diameter by means of a mandrel positioned between the rolls in the path of the metal.

Sometimes blanks are pierced in horizontal or vertical presses, and the shell thus obtained is then rolled into a pipe. This may be done in a common non-reversible two-high stand, in a continuous mill, or in a three-high mill.

The rolls of the two-high stand are provided with a number of round grooves to handle pipes of various sizes. Rolling of a shell into a pipe with the use of a mandrel is a hot process requiring two passes irrespective of the pipe wall thickness. As the stand is non-reversible, after the first pass the pipe is returned to the front side of the stand and tilted through 90°.

The continuous mill has nine stands with rolls set at 90°. The shell with the mandrel is forced by a pusher device between the rolls of the first stand. The pipe is rolled on the mandrel in all the stands at once. After rolling, the mandrel is removed.

Continuous mills of older design have seven alternating horizontal and vertical stands.

The rolls of a three-high mill are set in the housing at the apices of an equilateral triangle and rotate in the same direction. These mills are capable of producing pipes to very close tolerances.

The next operation is pipe finishing. Directly upon rolling, the pipes are subjected to reeling in a stand constructed similarly to the piercing stand. The purpose of reeling is to remedy the ovality, minimize the non-uniformity of wall thickness, improve the finish of both the external and internal surfaces of the pipes. Reeling is also performed on a mandrel, but the diameter of the latter is somewhat larger than the internal diameter of the rolled pipe. The pipe is somewhat expanded in the process of reeling.

Pipes are worked to final dimensions in a continuous sizing group which consists of several two-high stands.

The rolls have oval grooves, the ovality gradually diminishing, so that the final stand has a round pass. The stands are mounted on a common frame and alternately sloped 45° to both sides of the horizon.

The range of products of a mill can be extended by *reducing* the diameter of pipes. This is done in continuous reducing mills by working preheated pipes without the use of mandrels, which results in an increase in wall thickness. To prevent this, an adequate pull is applied to pipes in the process of reduction, the process being termed *stretch reducing*.

Pipes of small diameter with thin walls are obtained by cold rolling and drawing of hot-rolled pipes in special mills.

Manufacture of welded pipes. There are several methods for manufacturing welded pipes, but in all cases the starting blank is a steel strip or sheet.

One of the oldest techniques is furnace welding of pipes, lap and butt welding being both applicable.

When pipes are manufactured by the lap welding method, the preheated strip is rolled up in a welding bell. The pipe blank thus obtained is reheated and rolled on a mandrel in a two-high pipe mill. Welding is achieved due to the pressure in the focus of metal deformation. At present this method has fallen into disuse, as both its productivity and the quality of pipes are low.

In the butt welding method, the strip is also heated in a furnace, drawn through the welding bell in a manner that its edges come into contact with each other and weld together due to the pressure inside the welding bell.

At present, there are continuous lines for furnace welding of pipes. Before heating lengths of the strip are welded together (the back end of one being welded to the front end of the other so as to allow continuous operation). The strip is heated in tunnel-type furnaces of a great length and rolled in forming-and-welding multi-stand mills. The mill delivers an endless pipe which is cut by shears or saws. The pipe is then transferred to sizing mill where it is given its final dimensions.

Also widely used is the manufacture of pipes by electric welding. The starting material is a coiled band or sheet steel. The stock is straightened, welded into an endless strip and edge-trimmed. The strip is worked into pipes in forming mills. There are two main methods for forming pipes:

(1) The endless strip is rolled up into a coil until the side edges come into contact.

(2) The strip is wound spirally on a mandrel.

Pipes with a straight seam are obtained in the first case, and with a spiral seam, in the second. The advantage of the second method is that large-diameter pipes may be obtained from a relatively narrow strip.

The pipe stock thus formed is welded in a pipe-welding mill. Electric-welded pipes are manufactured by resistance welding, flash butt welding, submerged-arc welding, hydrogen-arc welding, and induction welding.

After welding, the weld is dressed, and the pipe is somewhat flattened in the process. The pipe is given final shape and dimensions in a sizing mill.

The pipes are cut as they come out of the mill with the aid of saws or special machining devices.

24.8. Rolling Defects

Defects in semi-finished and finished products may take their origin either in casting or in rolling.

Casting defects are non-metallic inclusions (slag, particles of refractory materials, etc.), double skins, cracks, teeming laps, tears, blow holes, excessive shrinkage cavities.

Non-metallic inclusions may have an adverse effect on both rolling (laminations and blocking of the strip in the mill) and on further processing, as well as on the use of a finished product. Thus, a metal may be torn at concentrations of non-metallic inclusions when it is subjected to deep drawing. In a finished product the non-metallic inclusions give rise to stress concentrations, and, hence, cracks.

Double skins, cracks, and tears are rolled out to great lengths and often cause rejection of finished products. Blow holes located deep inside an ingot are closed up in the process of rolling. When the blow holes are close to the surface, they open up in the process of heating and oxidize. As the metal is worked further, the defect is rolled off over a great length and necessitates the cleaning of the finished product surface. At times the defect may be so severe as to cause rejects.

A large shrinkage cavity requires additional cropping of the metal.

The defects of the rolling proper have their source in the failure to comply with the heating, rolling, cooling, and finishing schedules.

Defects from incorrect heating of ingots and blanks. *Burning* occurs when a metal is heated in an oxidizing atmosphere to a temperature close to its melting point. Oxygen penetrates inside the metal and oxidizes the grains, thus weakening the bonds between them. When rolled, a burned metal breaks into pieces.

Overheating, or prolonged holding of a metal at a high temperature, causes excessive grain growth and, therefore, impairs the plastic properties of the metal.

Scale is formed on the metal in the process of heating. If the heating procedure is correct, the loss in burning lies within one to three per cent, whereas failure to comply with the recommended heating schedule may result in a loss as high as five or six per cent. Besides, scale causes defects in further rolling.

Clinks are characteristic of high-carbon and alloy steels. Heat conductivity of these steels is low and a large temperature gradient between the surface and the internal layers occurs in the process of a rapid heating, causing internal stresses and, ultimately, failure of the metal.

Decarburization is the more pronounced, the higher is the temperature of heating and soaking. It is most undesirable for steels with a high carbon content.

All the above defects are caused by various departures from prescribed heating schedules. The main means of preventing the defects due to heating is strict compliance with the schedules of ingot and workpiece heating.

Defects due to rolling. The chief rolling defect is incorrect geometric shape and dimensions, the main cause being improper set-up of the mill.

Horizontal misalignment of grooves of a pass causes a rhombic cross section in blooms and blanks, undercut and twisting of the strip, distorted shape of sections (ovals instead of rounds, angles with wings of unequal thicknesses, etc.).

Vertical misalignment of the rolls produces blanks with unequal sides, twisted and cambered strip, deformed sections, sheet tapered in cross section.

Departures from prescribed reduction schedules affect the filling of the passes with metal: a pass may be either underfilled or overfilled. In both cases, the cross sections of both the blank and the finished product are distorted.

Pass underfilling leads to poor shaping of the horizontal angles of the blank, while rod iron and rounds are oval, with the horizontal dimension smaller than the vertical one, flange elements in sections are not shaped, etc.

Pass overfilling results in ribs or fins on both sides of blooms, blanks, and sections.

Incorrect set-up of inlet guides causes one-sided ribs and fins, undercut of strip, and wrong dimensioning of sections. *Undertilting or overtilting of strip* causes poor shaping of angles in blanks and square steel, distortion of shaped sections, and undercut of strip.

Incorrect shaping of the stock may also be due to wear of roll grooves and departures from rolling temperature schedules.

The lower is the rolling temperature, the worse are the plastic properties of a metal and the poorer is the longitudinal flow of a metal. Therefore, when metal is rolled at too low a temperature, the pass is overfilled, giving rise to ribs and fins.

If the passes are worn, and produce distorted sections, change over to another pass or replace the rolls.

Excessive wear of roll barrel in sheet rolling stands may result in an uneven thickness across the strip, the so-called *uneven gauge*. This may also be the result of rolling of chilled metal. At lower temperatures, metals become less ductile, so the pressure upon and, hence, the bending of the rolls increase, giving rise to uneven thickness across the sheet.

When thin sheet is rolled in continuous mills, the length of the strip may exceed 400 m. As the time of rolling of such a strip is one minute and its thickness is small, a temperature gradient arises between the front and back ends of the strip, which causes a longitudinal variation of thickness, termed *grow-back*.

Ridge buckle and camber of a sheet may be due to uneven transverse reduction of the blank. This irregularity may arise from incorrect roll design, uneven heating or wear of roll barrels, improper heating of strip, and considerable uneven gauge in a strip.

A number of other defects, such as backfins, wrinkles, burrs, and dog marks, may also arise during rolling.

Backfins resemble longitudinal cracks; they result from rolling of a strip marred by pinches, burrs, deep scratches.

Wrinkles are formed on rolling of a poorly shaped oval strip in a square pass.

Scores and *burrs* result from contact of strip with sharp metallic objects, frequently this being caused by a poor condition of the roll fittings.

Roll marks, observable at regular intervals along the length of the product, arise from damaged rolls or from pieces of steel adhering to the rolls.

Rolled-in metal chip is a frequent defect of cold-rolled sheet. Particles of metal that the moving strip chips off metallic objects fall on the strip surface, are rolled into it, and cause poor finish of the sheet. An efficient measure to prevent this defect is to manufacture parts in contact with the strip from soft materials, such as textolite or non-ferrous metals.

24.9. Economics of Rolling Plant

Rolling mill yield. The technically possible hourly yield of a mill is equal to 3,600 times the ratio of the blank mass to the rolling rate.

Rolling rate is the time interval between the feeding of two successive strips into the first stand.

If several passes are made in a single-stand mill, the rolling time for a strip is composed of the time required for each pass (productive time) and the time intervals between the passes.

The time between two successive engagements of strip depends either on the handling capacity of the first stand (in a blooming or a slabbing mill it is the time necessary to lift the top roll into the initial position) or on the capacities of other stands or units of the mill line.

In practice, the hourly production of a mill is lower than the technically possible yield because of the unavoidable delays which cause breaks in rolling. It is clear from the above that the larger the mass of a blank and the lower the rolling rate, the greater the hourly yield of a mill.

Determination of the yearly yield of a mill requires the knowledge of the hours of operation of the mill over a year.

Metal consumption. Rolling involves inevitable losses of metal, such as loss in burning in heating devices and during rolling, cropped ends, rejects and unfinished sections, loss of metal due to scarfing and mechanical peeling of blanks.

This is taken account of by a metal consumption factor, which is the ratio of the ingot mass to the mass of the finished product. This factor is greater than unity. The metal consumption factors for various types of mills are:

Blooming and slabbing mills	1.08-1.36
Section mills	1.05-1.10
Sheet mills (hot rolling)	1.03-1.27

The consumption of metal greatly depends on the grade of the rolled steel. Common grades of rimming steel have a lower factor, killed steels, a higher one. For example, the manufacture of one ton of finished sheet from Ст. 2 steel requires 1.05 tons of blanks (slabs).

This quantity of slabs, in turn, requires 1.155 tons of ingots; the total loss of metal for turning ingots into sheet reaches 155 kilograms.

CHAPTER 25

Drawing, Smith Forging, Die Forging, and Extrusion

25.1. Drawing

In the process of drawing, the metal is pulled through the aperture of the drawing tool, whose cross section gently tapers towards the exit.

Fig. 77a illustrates schematically the process of drawing.

Prior to drawing the leading tip of a rod is pointed, threaded up through the aperture, and gripped by a special pulling mechanism. Then the metal is drawn through the aperture, receiving the shape and dimensions of the latter.

The block-type draw benches are used for obtaining very long and thin products, most frequently, wire. The bench is provided with one or several blocks to wind the wire. Rotation of the blocks produces a tensile force which draws the wire through the aperture, or die hole.

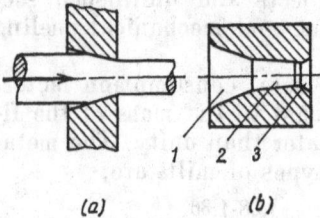

When the rod is passed through the hole of a single die, the unit is termed a single-block machine. When the rod is passed through the holes of two or more dies, the unit is known as a multi-die machine.

Wire is most usually drawn in slip-type multi-die machines. A die is set before each block. The wire passes through the first die and is wound on a block; then it slips down, is unwounded off the block, and enters the second die, after which it is wound on the second block, and so on.

Fig. 77. Drawing process (a) and die (b)

1 — approach; *2* — bearing; *3* — relief

In a slip-type bench the drawing speeds, i.e. the block rotation speeds, are interrelated with the diameters of the die holes.

Tungsten carbide dies (Fig. 77b) are used for wire drawing. The dies are set in die holders.

The leading end of the rod is pointed in special machines to enable its introduction (threading) into the die hole. Threading is a difficult operation, and, in order to obviate it, the ends of rod coils are butt-welded in electric welding machines so that only the end of the starting coil need be threaded, the process then continuing until there arises a necessity for replacing worn dies.

The work tool in pipe drawing is a ring die secured in a holder.

The starting material for drawing is a hot-rolled metal. The surface of the metal has a skin of hard scale, which causes a rapid wear of the drawing tool. This is why the scale must be removed by chemical or mechanical means before drawing. The mechanical method is seldom used as it is inefficient.

The chemical method consists in pickling the metal in solutions of sulfuric and hydrochloric acids.

In addition, the wire is subjected to sulling. The coils are placed in a pit and sprayed with atomized water. A yellowish oxide film (Fe_2O_3) formed on the surface of the wire acts as a lubricant during drawing.

Sulling is followed by liming. Wire coil is immersed in a boiling solution of lime to neutralize any traces of acids remaining on the surface of the metal.

Drawing involves friction, a considerable force being required to overcome it. Various lubricants, such as special soaps, a graphite-and-oil mixture and other compounds are used to minimize friction.

Drawing is a cold-working process, and it causes work hardening of the metal. Various kinds of heat treatment are employed after drawing to counter the ill effects of work hardening and to obtain the desired structure and mechanical properties of the metal.

Drawing is used to manufacture the following kinds of products:

(1) wire in diameters from 6.0 down to 0.4 mm, and sometimes even finer;

(2) tubes with walls thicknesses of 0.1-3.0 mm and in diameters down to 0.3 mm;

(3) various sections.

Items manufactured by drawing have a clean and smooth finish and an accurate cross section.

25.2. Smith Forging

Forging is a process of shaping metals by hammering.

Between the blows, the tool and the workpiece are moved in relation to one another. The processes of this type are called periodic as they are characterized by alternating work and idle strokes of the tool and by irregularity of the working effect over the length of the product due to a non-uniform deformation. Forging is practiced in two varieties, known as smith forging and die forging.

Hand forging is the oldest method of pressure working of metals. The metal is shaped on an anvil by striking it with a hand hammer or a sledge. At present, this method is used for the manufacture of small parts only.

Machine forging is performed with the aid of forging hammers (for small and medium-size forgings) or forging presses (for medium and large-size forgings). Machine forging produces large parts with small allowances for machining.

The starting material is either a standard rolled stock or ingots. Prior to forging, the metal is heated in furnaces

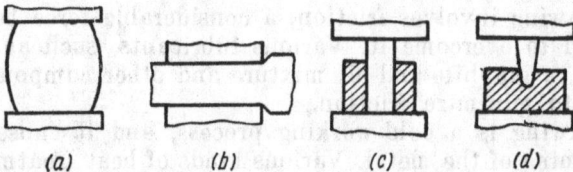

(a)   *(b)*   *(c)*   *(d)*

Fig. 78. Operations of smith forging

(a) upsetting; (b) drawing; (c) hollow forging; (d) punching

of various designs or in smithy forges. Commonly used are box and continuous furnaces; lately, induction furnaces have come into use.

Let us discuss the forging operations.

Upsetting (Fig. 78a). The whole of the blank is subjected to deformation. Its height diminishes, its cross section increases. By upsetting a blank on all sides, it may be worked back to its original shape, but the properties of the metal will have improved.

Drawing (Fig. 78b). Only part of the workpiece is in the deformation zone. After each blow the piece is moved forward, its length and breadth both increasing. Required dimensions and shape can be obtained by tilting the workpiece.

Hollow forging (Fig. 78c). The tool is smaller than the workpiece both in length and in width. In the process of working, the tool penetrates into the piece, piercing it through or making a depression in it.

Punching (Fig. 78*d*). The workpiece is sliced into two parts by a penetrating tool. If the penetration stops short of the complete separation of the blank in two, the operation is known as fullering.

Pressure welding. Two ends of a workpiece or of different workpieces are joined by pressure.

Besides the above operations, widely used are *bending*, *twisting*, and other processes.

Forge shops are chiefly equipped with power hammers, which work the metal at the expense of the kinetic energy of falling parts. They are simple in design and maintenance and reliable in operation.

Moving parts of the hammers are operated by compressed air or steam.

Compressed-air drop hammers are subdivided into single-acting and double-acting varieties depending on whether

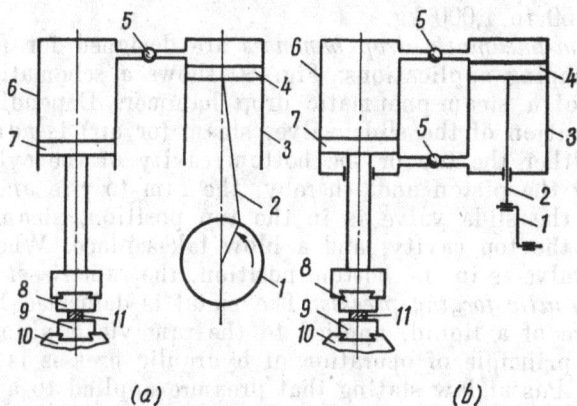

Fig. 79. Pneumatic hammers

(*a*) single-acting, (*b*) double-acting

1 — crankshaft; *2* — connecting rod; *3* — compressor cylinder; *4* — compressor piston; *5* — distributing cock; *6* — work cylinder; *7* — work cylinder piston; *8* — top block; *9* — sow block; *10* — anvil; *11* — workpiece

the compressed air acts from one side or from two sides of the piston (Fig. 79).

The single-acting type (Fig. 79*a*) operates as follows.

The compressor piston is driven by a crankshaft which, in turn, is rotated by an electric motor. As the compressor piston moves down, vacuum is created inside the work cylinder, atmospheric pressure causing the piston of the work cylinder to rise. The compressor piston then rises, producing a pressure both inside the compressor cylinder and inside the work cylinder, so that the work piston goes down to strike a blow, and so on. A distributing cock allows idle running by isolating the work cylinder from the compressor or delivery of separate blows. Hammers of this type are most frequently used with ram masses of 50 kg.

Somewhat more complicated is the operation of a double-acting hammer (Fig. 79*b*). Air is supplied to the top part of the work cylinder when a blow is to be effected, and to the bottom part underneath the work piston to lift the ram.

These types of hammers are manufactured with ram masses of 50 to 1,000 kg.

Steam-pneumatic drop hammers are designed for forging or stamping applications. Fig. 80 shows a schematic diagram of a steam-pneumatic drop hammer. Depending on the position of the slide valve, steam (or air) is admitted into either the top or the bottom cavity of the cylinder, forcing the piston and, thereby, the ram to rise and fall. When the slide valve is in the top position, steam (air) enters the top cavity, and a blow takes place. When the slide valve is in the bottom position, the ram rises.

Hydraulic forging presses. The metal is deformed by the pressure of a liquid, applied to the ram via a plunger.

The principle of operation of hydraulic presses is based on the Pascal law stating that pressure applied to a given area of a fluid enclosed in a vessel is transmitted undiminished to every equal area of the vessel.

Fig. 81 shows a schematic diagram of a hydraulic press. Water fed under a high pressure into the work cylinder *1* drives the piston down. Ram *3* rigidly linked to the piston also goes down and acts on the workpiece. The ram is raised by lifting cylinders whose pistons are connected to the crosspiece. Water is supplied to the bottom part of the lifting cylinders and forces the piston to rise together with the ram of the press.

Hand forging is the oldest method of pressure working of metals. The metal is shaped on an anvil by striking it with a hand hammer or a sledge. At present, this method is used for the manufacture of small parts only.

Machine forging is performed with the aid of forging hammers (for small and medium-size forgings) or forging presses (for medium and large-size forgings). Machine forging produces large parts with small allowances for machining.

The starting material is either a standard rolled stock or ingots. Prior to forging, the metal is heated in furnaces

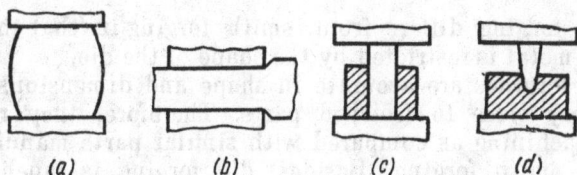

(a)   *(b)*   *(c)*   *(d)*

Fig. 78. Operations of smith forging

(a) upsetting; (b) drawing; (c) hollow forging; (d) punching

of various designs or in smithy forges. Commonly used are box and continuous furnaces; lately, induction furnaces have come into use.

Let us discuss the forging operations.

Upsetting (Fig. 78a). The whole of the blank is subjected to deformation. Its height diminishes, its cross section increases. By upsetting a blank on all sides, it may be worked back to its original shape, but the properties of the metal will have improved.

Drawing (Fig. 78b). Only part of the workpiece is in the deformation zone. After each blow the piece is moved forward, its length and breadth both increasing. Required dimensions and shape can be obtained by tilting the workpiece.

Hollow forging (Fig. 78c). The tool is smaller than the workpiece both in length and in width. In the process of working, the tool penetrates into the piece, piercing it through or making a depression in it.

Blanks are cut and trimmed with the aid of anvil chisels of various shapes. Holes are pierced with punches.

Power-hammer dies and swages allow forging to more accurate dimensions, thus ensuring minimum machining allowances.

Various mechanisms and devices, such as cranes, tilters, chucks, grips, tongs, crowbars, and levers, are employed to handle and manipulate forgings whose mass may reach tens of tons.

25.3. Die Forging

Die forging differs from smith forging in that the flow of the metal is restricted by the shape of the die.

Die forgings are accurate in shape and dimensions, they are very close to finished parts. Therefore, they require less machining as compared with similar parts manufactured by smith forging. Besides, die forging is much more productive than smith forging, and it is employed chiefly in mass and large-series production. In small production runs, die forging is utilized less frequently as the cost of dies is fairly high.

Die forging is essentially as follows: the blank is heated and placed into the lower half of the die; the upper half of the die, which also has an impression, goes down and compresses the blank, causing the metal to flow and fill the die.

In case of an intricate part, the blank is preforged by smith forging or in a preforming impression of the die in case of multi-die forging. If necessary, the finished forging is heat treated and machined.

Die forging equipment and tools will be discussed below. Let us note here that the cavity of a given shape, or the *impression*, is formed by the two halves of the die. Placing of the blank and removal of the die-forged part take place with the die open.

Good results are obtained when the mass of the blank is somewhat larger than the required mass of the forging. In this case, the die impression is well filled with the metal, the excess material being forced out to form a flash (Fig. 82*a*).

If flash formation is to be avoided, the volume of the blank must be exactly equal to that of the forging (Fig. 82b).

Die forging equipment. Steam-pneumatic and mechanical drop hammers, mechanical and hydraulic presses, and horizontal forging machines are used for die forging.

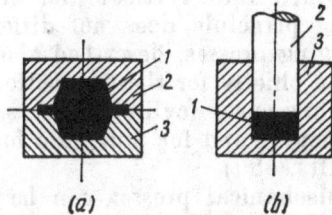

(a)   *(b)*

Fig. 82. Die forging methods

(a) flash formation; (b) flash not formed

1 — blank; *2* — top die block; *3* — bottom die block

A steam-pneumatic hammer for die forging has little difference from similar hammers destined for smith forging applications, which have been described above. An essential requirement upon the hammer is that its blows should be accurate.

Among the mechanical hammers is the *board-drop hammer* (Fig. 83).

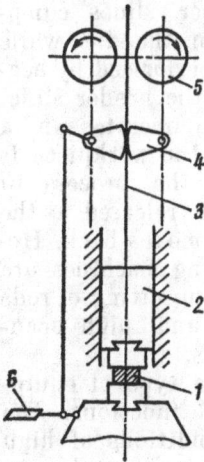

Fig. 83.   Board-drop hammer

1 — blank; *2* — ram; *3* — board; *4* — clamps; *5* — rolls; *6* — treadle

This type of hammers operates as follows. Rolls *5* rotate continuously in opposite directions and lift, due to friction, board *3* together with ram *2* into a striking position, clamps *4* letting the board pass as it is raised. As soon as the board reaches its top position the rolls move aside, and the board is gripped by the clamps. Depressing treadle *6* separates the clamps, and the board with the ram falls, striking blank *1*. At this moment the rolls close in on the board which is thus again moved upward, and the cycle repeats.

Hydraulic die-forging presses produce far greater forces than the forging presses. Presses capable of providing a force of 800 MN have been put into service recently.

Hydraulic die-forging presses are

divided into vertical and horizontal types. Their operating principle does not differ from that of the hydraulic
forging presses, described above. Today, these presses are
used chiefly for the production of ferrous and non-ferrous
hollow parts (cylinder liners, gun shells, etc.), for sheet
stamping and for multi-die forging of intricate parts (crankshafts, etc.)

Mechanical presses can be exemplified by *screw presses*
in which metal is deformed by the force produced by a rotating screw (or nut). Screw presses develop relatively small
forces and are used for the manufacture of small forgings
(bolts, nuts, etc.) and non-ferrous parts, as well as for such
operations as straightening, bending, sizing, and die stamping. A schematic diagram of a horizontal forging machine is
shown in Fig. 84. The crankshaft of the machine is rotated
by an electric motor. The re-
ciprocating motion is transmitted via a connecting rod
to header slide 2 and via a
leverage to gripper slides 3.

The gripper slides comp-
ress rod 1 in the die while
upsetting is performed by header 4 set in the header slide.
The machine operates in a
manner that the workpiece is
clamped in the process of
upsetting and released as the
header slide moves back. Horizontal forging machines are
used for hot upsetting of rods
and pipes. They ensure high productivity and high accuracy of parts, and operate without impacts.

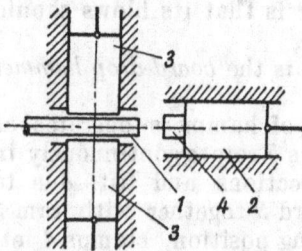

Fig. 84. Horizontal for-
ging machine

1 — rod; *2* — header slide; *3* —
gripper slides; *4* — header

Dies. The die design is governed by the type of equipment to operate them and their intended functions. The
dies are subjected to severe operating conditions of high
temperature and high pressure of the workpiece metal upon
the die. Therefore, dies are manufactured from special
steels and subjected to heat-treatment.

Commonly, dies are composed of two parts with engrav-
ings. When put together, the two halves form a die, the

Hand forging is the oldest method of pressure working of metals. The metal is shaped on an anvil by striking it with a hand hammer or a sledge. At present, this method is used for the manufacture of small parts only.

Machine forging is performed with the aid of forging hammers (for small and medium-size forgings) or forging presses (for medium and large-size forgings). Machine forging produces large parts with small allowances for machining.

The starting material is either a standard rolled stock or ingots. Prior to forging, the metal is heated in furnaces

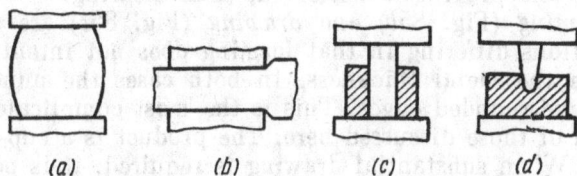

(a)   *(b)*   *(c)*   *(d)*

Fig. 78. Operations of smith forging

(a) upsetting; (b) drawing; (c) hollow forging; (d) punching

of various designs or in smithy forges. Commonly used are box and continuous furnaces; lately, induction furnaces have come into use.

Let us discuss the forging operations.

Upsetting (Fig. 78a). The whole of the blank is subjected to deformation. Its height diminishes, its cross section increases. By upsetting a blank on all sides, it may be worked back to its original shape, but the properties of the metal will have improved.

Drawing (Fig. 78b). Only part of the workpiece is in the deformation zone. After each blow the piece is moved forward, its length and breadth both increasing. Required dimensions and shape can be obtained by tilting the workpiece.

Hollow forging (Fig. 78c). The tool is smaller than the workpiece both in length and in width. In the process of working, the tool penetrates into the piece, piercing it through or making a depression in it.

Fig. 85 shows the principal pressworking operations. The tools used are the punch, the die, and the pressure pad, the latter being sometimes omitted.

Let us review in brief the pressworking operations.

Blanking (Fig. 85a) is used for the manufacture, from sheet metal, of round or differently-shaped parts which usually serve as blanks for the fabrication of finished items.

Punching of holes of different configurations in flat and relatively thin parts is similar to blanking. The punch and die have sharp edges. The method is much more productive and provides a better cut than shearing.

Forming (Fig. 85b) and *drawing* (Fig. 85c) are similar operations differing in that forming does not intentionally reduce the metal thickness. In both cases the punch and die have rounded edges. This is the most complicated operation of those discussed here. The product is a cup-shaped item. When substantial drawing is required, it is performed in several steps with intermediate anneals.

Bending (Fig. 85d) is used most frequently on soft steels as the workpiece springs back.

Pressworking uses crank, friction, and excentric-type presses.

25.5. Extrusion of Solid and Hollow Sections

A great variety of solid and hollow sections is manufactured by extrusion, the principle of this process being schematically illustrated in Fig. 86.

Heated billet 5 is placed in container 1, one side of which is closed with die 4 with a hole, and on the other side there are dummy block 3 and ram 6. The ram acts upon the dummy block, and, through it, upon the metal which is thus forced through the orifice in the die. The cross section of the outflowing metal is shaped by the die orifice. Fig. 86a illustrates schematically the method of direct extrusion in which the metal flows in the direction of the plunger motion.

In the method of indirect extrusion, the orifice is made in the dummy block, so that the metal flows in the direction opposite to that of the plunger movement (Fig. 86b).

Solid sections are manufactured usually by the indirect method, and hollow sections, by the direct method. When extruding hollow sections, the billet is first pierced by a mandrel whose end reaches the hole and forms with it an annular opening. As the ram and the dummy block move

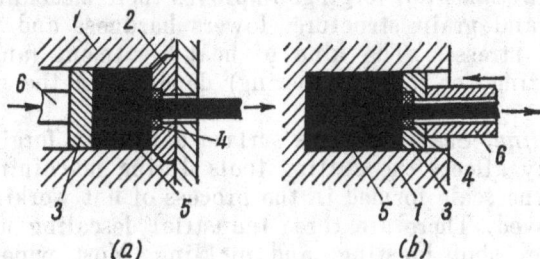

Fig. 86. Process of extrusion

(a) direct method; (b) indirect method
1 — container; *2* — die holder; *3* — dummy block; *4* — die; *5* — billet metal; *6* — ram

forward, the metal flows through the annular opening to form a tube or another shape of hollow section. Extrusion is performed on hydraulic or mechanical horizontal and vertical presses.

The billets for the process are either ferrous rolled stock or ingots of non-ferrous metals and alloys. Before extrusion, the billets are heated to a certain temperature which depends on the composition of the metal or alloy. The container, die, dummy block, and mandrel are also heated. Extrusion is employed for the manufacture of items of intricate cross sections to close tolerances and with a smooth finish.

25.6. Operations Subsequent on Forging

Forgings are subjected to the following operations: straightening, heat treatment, descaling, and sizing.

Straightening is necessary when forgings have been warped during manufacture, handling, heat treatment, trimming, etc. Deformation is revealed by a tracing-up plate or by other devices. When this deformation exceeds prescribed tolerances, straightening is the remedy. Hot straightening is performed in the basic die, on a trimming press or in special straightening equipment. Cold straightening is intended to correct warpage resulting from heat treatment and is performed with the aid of board-drop hammers.

Heat treatment of forgings improves their mechanical properties and grain structure, lowers hardness and removes residual stresses. The kind of heat treatment (annealing, normalizing, or high tempering) depends on the grade of steel.

Descaling. Scale conceals surface defects in forgings and adversely affects the cutting tools during machining. This is why the scale formed in the process of hot working must be removed. There are three industrial descaling methods: tumbling, shot blasting, and pickling. Most widely used today are the former two methods.

Tumbling is applicable to small forgings only. These are placed in a barrel whose rotation causes the forgings to impinge upon one another and thus break off the scale.

In shot blasting, a high-speed jet of cast-iron shot (1.5-2 mm in diameter) strikes over the forging and knocks the scale off.

Pickling was formerly used on a wide scale, but owing to low efficiency and harmful conditions of work is now seldom used.

Sizing which consists in a small compression of the piece improves the dimensional accuracy of forgings.

25.7. Defects of Items Manufactured by Drawing, Forging, and Extrusion

In pieces made by plastic working of metals, defects may be due either to a poor quality of the starting material (blank) or to flaws arising in preparatory operations and directly in the process of drawing, forging, and extrusion.

Blank defects include seams, double skins, non-metallic inclusions, scratches, hair cracks, laps; these defects can be traced down to rolling.

Prior to plastic working, metal is usually heated. Characteristic defects arising during the operation are overheat and burning caused by poor heating practice; here also belongs intensive scaling.

The principal defect in plastic working proper is departure from prescribed dimensions. This type of defects may result from the wear of the work tool, deviation from reduction schedules, improper dimensions of the starting blank, poor mounting and set-up of the tool and equipment. A usual kind of defects is that of mechanical properties and grain structure of the metal.

Besides, there are defects characteristic of every particular method of plastic working. For example, often encountered in drawing are ovality of cross section and local necking. Defects characteristic of forging are forged-in scale, incomplete filling of the die impression, forming of crowds and shuts, internal and external cracks, tears and burrs, dents, etc.

Strict compliance to heating, reduction, and heat-treatment schedules, i.e., good practice at all production stages, and the operation of serviceable and properly set up equipment are essential to the manufacture of high-quality products.

CHAPTER 26

Quality Inspection of Finished Items

Items produced by plastic working of metals are manufactured to standards and specifications.

Respective standards specify the dimensions and tolerances for finished products. Depending on the kind of products, the specifications may include additionally the cross-sectional area and the mass of a metre of section length, the length of strip to be supplied, the tolerances on the length of rolled stock, while standards for beams, channels, etc. may specify the modulus of section, the moment of inertia, the radius of gyration, etc.

Standards and specifications may also cover the chemical composition of steel, its mechanical properties, finish, macro- and micro-structure, as well as rules of acceptance of the products, testing methods, identification marking, etc.

When non-standard items are to be manufactured, specifications are drawn up listing the customer's requirements as to the dimensions and tolerances of the finished product, its properties, and the inspection methods.

Measuring instruments. The *vernier caliper* is used to measure the outside and inside dimensions of parts. The whole number of millimetres is read off the bar scale, while fractions of a millimetre are indicated by the vernier scale. Measurements can be made to an accuracy of 0.1, 0.05 or 0.02 mm, depending on the value of the vernier division.

The *vernier depth gauge* is intended for measuring the depth of blind holes, grooves, shoulders, recesses, etc. A depth gauge resembles a vernier caliper in appearance.

Outside calipers are used to measure outside, and the *inside calipers*, the inside dimensions. They differ only in the shape of leg ends. The distance between the caliper legs is measured by means of a rule.

Micrometers serve to measure the outside dimensions to within 0.01 or even 0.002 mm. The measurement range is usually 0 to 25 mm and 25 to 50 mm.

A *dial indicator* is intended for measuring departures from specified dimensions. The measurement ranges are 0-2, 0-3, 0-5, and 0-10 mm.

The above instruments are used to check the dimensions of most various items. In addition, there are instruments and devices capable of checking the dimensions of one specific type of products.

Thus, items manufactured by plastic working of a metal are often checked with *limit gauges*. A typical limit gauge has two apertures cut to the shape of the rolled section to be checked. One of the apertures corresponds to the high limit, and the other to the low limit of the tolerance range. When a section passes freely through the former aperture (the "go" side) and does not pass through the latter aperture (the "no-go" side), the item is within the tolerance range.

Forgings are also checked with the use of limit gauges. Gauges used for checking shafts are called *snap gauges*, those for checking holes, *plug gauges*.

Appearance inspection. A finished product is visually inspected at specially equipped and well illuminated areas. Detected surface flaws must be carefully eliminated.

Surface finish is of particular importance to some types of products, such as sheet metal. Standards usually subdivide sheet metal into a number of surface finish grades.

Metallographic examinations. Metallographic examinations are aimed at inspecting the macro- and micro-structures of metals. Macrostructure is determined at small magnifications or by a naked eye, whereas microstructure is viewed in a microscope on a specially prepared specimen of the metal.

Metallographic examinations afford a means to assess the degree of liquation of such elements as carbon, sulfur, and phosphorus, the presence and character of non-metallic inclusions, the grain size, the distribution of structural components, the extent of decarburization, etc.

Non-destructive testing. The methods of non-destructive testing have been devised chiefly for detecting internal defects of items.

No method of visual inspection can detect internal defects that do not open upon the surface. These defects show up either in the process of further working or in service where they may lead to equipment failures and accidents.

Ultrasonic flaw detection based on the capability of ultrasonic waves to reflect from defective area is a technique widely used in inspection of rolled stock and forgings. This method has many advantages: the examination is practically instantaneous and poses no limitations on the size of parts or workpieces under test. The checks are carried out with the aid of a special instrument, termed the ultrasonic flaw detector.

Radiographic inspection for detecting internal defects is based on the different penetrating ability of X-rays in materials of different densities. If lying in the path of the rays are voids, cracks, non-metallic inclusions, and the like, the intensity of the passing rays changes, and the configuration of the defect appears on a screen or a photographic plate.

X-ray equipment is very expensive, and this has stimulated in recent years the use of radioisotope (gamma-ray and beta-ray) flaw detectors. Gamma and beta rays are absorbed differently by materials of different densities.

The magnetic method is used to detect minute defects,

such as small cracks, seams, blow holes, on the surface of metals. The method is based essentially on that a metallic powder forms characteristic patterns at sites of defects in a magnetized item.

CHAPTER 27

Mechanization and Automation of Processes of Plastic Working of Metals

Most of the operations at rolling plants, such as placing of ingots and blanks into soaking pits and heating furnaces, delivery of blanks to mills, rolling proper, cutting, handling and shipment of finished stock are at present mechanized. The operation of quite a lot of mechanisms and groups of mechanisms in rolling mills is automatic. Thus, blooming and slabbing mills have automatic receiving and main roll tables, screwdown arrangements, and main roll drives. In section mills, roller tables, tilting devices, mechanisms for ejecting strip onto cooling banks and for transporting it along cooling banks operate automatically. The operation of all these mechanisms is controlled by photocell relays or other transducers.

Both hot- and cold-rolling continuous mills are equipped with automatic systems for continuous control of strip thickness and width. The reels and flying shears are also operated automatically.

Development work is now under way to computerize the rolling mill control. Computers will control the operation of a rolling mill, i.e., set up the mill, control the dimensions of the rolled strip, and maintain a correct temperature schedule.

Until recently, only the main operations, and not all of them at that, were mechanized at forging plants. Auxiliary operations were performed manually.

The first to be mechanized were the operations that could not be performed manually, *viz.*, handling and working of large-size forgings and workpieces.

Blanks and ingots are charged into and discharged from furnaces with the aid of travelling cranes, lifts of various types, charging machines, and manipulators.

Transportation of a workpiece to a hammer or press, its tilting in the process of forging, and other operations are performed by means of manipulators.

In the forging of small parts, mechanization has little beneficial effect on the productivity of labour, but improves the working conditions greatly.

Automation in smith forging lags behind as compared to rolling, only the operation of heating furnaces and the control of blank dimensions being automatic.

Die forging work has been automated on a much more extensive scale. Automatic die-forging lines are in wide use. They are very efficient, turn out products of a high quality, and may be manned by a very limited number of workers or even by a single operator.

REVIEW QUESTIONS

1. What is *plastic deformation of metals*?
2. What is *section*?
3. What furnaces are used for heating metals?
4. How are rolling mills classified?
5. What units make up the main train of a rolling mill?
6. Name the auxiliary mechanisms of rolling mills.
7. What is *roll pass design*?
8. What pass design series are in use?
9. Name the principal operations in the manufacture of sheet and section stock.
10. Name the tube manufacturing methods.
11. Describe the defects of rolled stock.
12. What type of equipment is used for drawing metal?
13. Name the principal operations in drawing.
14. What forging operations are known?
15. What equipment and tools are used in forging?
16. Name the pressworking operations.
17. What are the methods of extruding?
18. In what manner is scale removed from the surface of forgings?
19. What defects may occur in drawing, forging and extrusion?
20. Describe the instruments for size measurement.
21. Speak on the methods of non-destructive testing.
22. What is *mechanization and automation* of production processes. and what is the difference between them?

SECTION VI • FUNDAMENTALS OF HEAT TREATMENT OF METALS

It has already been said that the properties of steel are governed by its structure and chemical composition. The aim of heat treatment of steel is to control its properties by changing its structure through heating and cooling.

CHAPTER 28

Purpose and Kinds of Heat Treatment

28.1. The Essence of Heat Treatment

Heat treatment is very important for present-day engineering. Practically all steels are subjected to heat treatment. Heat treatment can, for example, cause the following changes:

(1) soften steel to facilitate its machining;

(2) enhance hardness and wear resistance of a steel part operating in contact with other materials (tools, gears, rails, etc.);

(3) reduce the grain size in steel and thus impart additional toughness to it;

(4) increase the grain size in steel in order to give it some special properties;

(5) improve the magnetic properties of steel;

(6) ensure maximum plasticity of sheet steel intended for stamping;

(7) enhance corrosion resistance and heat resistance of steel by saturating its surface with other elements.

Heat treatment is possible because solid steel, as evidenced by the iron-carbon equilibrium diagram (Fig. 8), is the site of structural transformations in the process of heating and cooling.

When, for example, a steel is heated to 1,000°C, the following changes occur in its structure.

Alpha-iron changes to gamma-iron, and the cubic lattice transforms from body-centered into face-centered. Atoms of carbon are redistributed as the face-centered lattice is capable of dissolving all of the available carbon. The steel, therefore, acquires a single-phase austenitic structure, even though at room temperature its structure might have been composed of, say, ferrite and pearlite or pearlite and cementite.

From the iron-carbon equilibrium diagram it may well be expected that reverse processes will take place in the steel during cooling. And this actually occurs in the process of a relatively slow cooling: the structure is restored in its original state, and the recrystallization in the solid state affects the grain size only. This heat treatment operation is known as *annealing*.

When a steel is cooled rapidly, the transformations in it follow a different pattern. Like in the case of slow cooling, gamma-iron transforms into alpha-iron, since a polymorphic transformation cannot be retarded by rapid cooling. As regards carbon, the distribution of its atoms has no time to change and thus it remains in the lattice of the alpha-iron.

However, alpha-iron is capable of dissolving a maximum of 0.02 per cent carbon, whereas the high limit for carbon in steel is 2 per cent. In consequence, the concentration of carbon in alpha-iron will not be a normal and equilibrium one (as in ferrite), but will be much higher. This supersaturated solid solution of carbon in alpha-iron is termed *martensite* (so named after the metallurgist A. Martens), and the process of heating and rapid cooling is called *quenching* or *hardening*.

Thus, the structure of a steel can be altered by a combination of heating and cooling at various rates.

If a martensitic-hardened steel is heated again, the atoms of carbon acquire considerable energy and mobility and may again break away from the alpha-iron lattice to form cementite, with the effect that the steel will again acquire a ferrite-pearlitic structure, in conformance to the equilibrium diagram. Heating of hardened steel followed by slow cooling is termed *tempering*.

Besides, steels can be subjected to case-hardening which is essentially the saturation of the metallic surface with a certain element, such as carbon (carburizing), nitrogen (nitriding), aluminium (calorizing), etc., so as to provide the surface with specific properties, e.g., wear resistance, hardness, corrosion resistance, etc.

28.2. Annealing

From what has been said in the previous paragraph it follows that annealing is the type of heat treatment that results in equilibrium structures appearing on disintegration of austenite during cooling. Like other kinds of heat treatment, annealing consists in heating the metal to a given temperature, holding it at that temperature and subsequently cooling it under predetermined conditions.

The aim of annealing is to remove internal stresses, improve structural homogeneity and machinability, reduce hardness, and enhance plasticity.

Steel is usually annealed in the form of forgings, rolled stock, ingots, and castings. There are various kinds of annealing treatment, which differ in purpose and temperature schedules; the most widely employed ones are discussed below (see Fig. 87).

Homogenizing is heating to 1,000-1,100°C, prolonged holding at this temperature, and slow cooling in the furnace. Homogenizing is most often applied to large-size castings, forgings, and ingots.

Prolonged heating ensures a homogeneous chemical composition of the metal, but it causes a considerable coarsening of the grain. Fine-grain structures may be obtained by subsequent plastic working (forging) or by special heat treatment.

Full annealing consists in heating to a temperature 30-50 degrees (Centigrade) above the A_3 point (see Chapter 3), holding and slow cooling. In the process of heating the ferrite-pearlitic structure transforms into austenitic, whereas during cooling the austenite disintegrates to a ferrite-pearlitic mixture, i.e., *recrystallization* of the steel occurs. The purpose of recrystallization is to correct the coarse-grain structure of as-cast or as-forged steel and to relieve internal stresses.

A variety of full annealing is *normalizing* whose feature is cooling in the air. The heating temperature for normaliz-

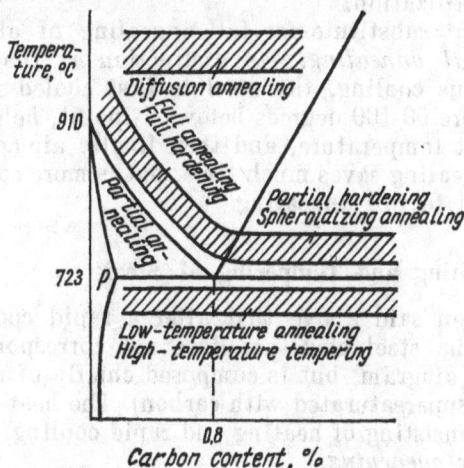

Fig. 87. Part of iron-carbon equilibrium diagram, corresponding to steels. Cross-hatched areas indicate the heating temperatures for various kinds of heat treatment

ing is by 30-100 degrees (Centigrade) above the A_3 point; the final structure has a finer grain.

In hypoeutectoid steels (i.e., in steels with up to 0.8 per cent carbon), full annealing and normalizing ensure practically the same properties, therefore, the cheaper and quicker process of normalizing is usually chosen.

If the initial grain structure is satisfactory and there is no need for a full recrystallization of both ferrite and pearlite, partial annealing is applied, which results in the recrystallization of pearlite only. In this case the annealing temperature is by 30-50 degrees above A_1 point.

The purpose of partial annealing for a hypoeutectoid steel is some improvement of its structure and machinability and relieving of internal stresses.

Partial annealing of hypereutectoid steels (*spheroidizing*) results in nodular pearlite instead of the flaky pearlite. This structure, in which cementite is in the form of fine spherical grains, has better machinability and lower tendency to overheat in the process of hardening.

Low annealing is used to relieve internal stresses without recrystallization.

A frequent substitute for full annealing of alloy steels is *isothermal annealing*. Its distinction is that, instead of continuous cooling, the steel is first cooled rapidly to a temperature 50-100 degrees below A_1 point, held for some time at that temperature, and then finally air cooled. Isothermal annealing saves much time, but is more complicated as compared to full annealing.

28.3. Hardening and Tempering of Steel

It has been said above that after a rapid cooling of a hot steel, the steel structure does not correspond to the equilibrium diagram, but is composed chiefly of martensite (alpha-iron supersaturated with carbon). The heat-treatment procedure consisting of heating and rapid cooling, is termed *hardening* or *quenching*.

Martensite is a very hard and brittle structural component, therefore, hardened steel has a maximum hardness, equal to 60-63 Rockwell units. The microscope shows an acicular structure of martensite (Fig. 88).

As in the case of annealing, the hardening temperature is chosen according to the iron-carbon equilibrium diagram on the basis of the carbon content in the steel.

Full hardening. Hypoeutectoid steels (with up to 0.8 per cent carbon) are fully hardened by cooling down from a temperature 30-50 degrees above A_3 point (see Chapter 3). At this temperature the structure of steel consists of austenite, and after hardening, of martensite. If a steel is hardened from a lower temperature, some amount of ferrite will remain unconverted after cooling, and the hardness of steel will be lower. Therefore, the aim of hardening can be attained only by heating the steel to a temperature at which it consists of austenite only.

However, no full hardening is required for hypereutectoid steels with more than 0.8 per cent carbon. At a temperature somewhat above A_1 point, their structure is composed of austenite and cementite. If such a steel is hardened, it will consist of martensite and cementite. These two constituents are very hard, therefore, the purpose of hardening

Fig. 88. Acicular martensite in hardened grade 45 steel (×300)

will be attained. This is why hypereutectoid steels are subjected to *partial hardening*.

The quenching medium most often is water or oil, both ensuring sufficiently high cooling rates for austenite to convert into martensite. Water solutions of salts, molten salts and lead may also serve as quenching media. Some alloy steels are quenched in the air.

Rapid cooling causes heavy internal stresses that may result in a warpage or even cracking of the part.

Internal stresses may be lessened, especially in alloy steels, by appropriate hardening techniques.

Austempering is effected by quenching the steel in a medium (e.g., a molten salt) heated to 200-500 °C. It is applicable to tools, gears, bearings.

Interrupted quenching is effected by cooling the metal successively in two media, for example, in water and then in oil, or in water, and then in the air; it is applied to intricate and heavy parts.

A number of other hardening techniques, such as chill quenching, temper quenching, are also practiced.

A hardened steel is not only very hard, but also very brittle, this being an undesirable property. Besides, a high hardness poses problems in final machining. To counter this, a hardened steel is subjected to *tempering*, that is, heating followed by air or water cooling.

Tempered steel is less hard and brittle, martensite having disintegrated into structures similar to pearlite, but of a finer constitution, such as *troostite* (so named after the French scientist G. Troost) *or sorbite* (named after the British scholar H. Sorby). In addition, tempering relieves internal stresses.

A low (150-200°C), medium (300-450°C), or high (450-650°C) tempering is used depending on the required combination of hardness and toughness.

28.4. Case-hardening of Steel

Case-hardening is a treatment of a thin surface layer (case) of a steel part with the view to imparting it special properties, such as wear resistance (carburizing or nitriding), resistance to high-temperature oxidation (calorizing), etc.

In some applications there is no point in endowing the bulk of metal with special properties, as it is sufficient to alter the structure and properties of the surface layer only.

Case-hardening is effected by placing a part into a solid, liquid, or gaseous medium containing atoms of the element capable of communicating the required properties to the surface layer. At high temperatures, the atoms of the saturating elements (carbon, nitrogen, sulfur, aluminium, etc.) penetrate in the part and may react chemically with iron and other constituents of steel, forming new structural components; the properties of the part's core remain unaltered.

Carburizing, i.e., saturation of steel surface with carbon, is one of the most common types of case-hardening in mechanical engineering.

The part to be carburized is placed in a furnace filled with a mixture of carbon monoxide and hydrocarbons (gas carburizing), or in a container filled with charcoal and catalyzers (solid carburizing). At 900-950°C carbon monoxide and other carbon-bearing substances decompose to give atomic carbon.

The nascent carbon saturates the surface of the part and forms a high-carbon layer of a thickness ranging from one to a few millimetres, depending on the duration of holding at the carburizing temperature.

Subjected to carburizing are low-carbon steels (up to 0.3 per cent carbon). The core of the part retains good toughness, even though the surface layer becomes very hard and brittle, as the carburized layer has harder structural constituents (pearlite and cementite) than the starting ferrite-pearlite structure.

In order to obtain a maximum effect out of carburizing, the process is complemented by a heat treatment improving the structures of both the surface layer and the core: a single or double quenching is followed by a low tempering. As the core and the carburized layer have different carbon contents, the temperature of hardening aimed at obtaining a martensitic structure will naturally be lower than that of hardening aimed at improving the core strength.

Of the other kinds of case-hardening techniques widely used are *nitriding* which enhances hardness, wear resistance and corrosion resistance of steel, *cyaniding*, or simultaneous saturation of steel surface with carbon and nitrogen whose purpose is to increase hardness and wear resistance, and *calorizing* whose object is to improve the corrosion resistance of manufactured items.

28.5. Heat Treatment Equipment

Steel is usually heat-treated at machine-building plants, heat treatment being part of the manufacturing process. In a number of cases, however, heat treatment takes place at a metallurgical plant, where it is included in the metal production process.

Heat-treated at iron-and-steel works are usually steel sections (rod iron, reinforcement iron), sheet and plate steel, rails. The main purpose in these cases is to increase the strength of the metal.

Rolling of steel is commonly completed at temperatures of 900-1,000 °C, after which the metal is left to cool in the air, i.e., is practically normalized from the rolling temperature.

At a machine-building plant the aims of heat treatment are more diversified. They include, for example, annealing to improve machinability, tempering to relieve internal stresses, temper hardening to increase strength and obtain adequate toughness.

Heat treatment is effected with the aid of special equipment either at a heat-treatment shop or at specialized departments of rolling, forging and other production shops.

Heat-treatment equipment may be subdivided into main (furnaces, quenching tanks, etc.) and auxiliary (pickling and rinsing tanks, cleaning devices, etc.) devices.

A heat-treating furnace is a complicated unit that must ensure a uniform and sufficiently rapid heating of steel, a convenient charging and discharging of parts, a dependable temperature control, and a prescribed composition of the furnace atmosphere preventing excessive oxidation or burnout of carbon from the surface of steel.

There are heat-treating furnaces of various types to suit the various kinds of heat treatment, type of fuel, throughput capacity, and items treated.

For small parts most often used are *box furnaces* (Fig. 89). The parts are charged on hearth plate *1* and discharged from it through door *2*. Spiral or ribbon heating resistors *4* are secured to lining *3* throughout the length of the heating chamber. A thermocouple is introduced into the heating chamber through the furnace shell and lining to control the temperature.

Instead of the wire heaters from heat-resistant alloys, which provide a maximum temperature of 1,100 °C, the furnace may be heated also with silicon-carbide heating elements capable of temperatures up to 1,300 °C.

Box furnaces are relatively simple in design and operation, but they have a limited productivity, particularly when heating large-size ingots and forgings.

A more efficient design is that of *conveyor furnaces* used for heat treatment of ingots, forgings, sheet, machine parts, etc. They feature the possibility of moving the parts in the process of heating. The items are put on a conveyor or a roll table, sometimes on special pans, and advanced in the furnace toward the exit end. The length of such a furnace may reach tens of metres.

The furnace length may be divided into several zones with different temperatures, this ensuring a greater accuracy of control of the heating rate and heating of all the parts under more uniform conditions.

There are also furnaces specially suited to heat treatment of special types of products. *Bell furnaces* for annealing sheet steel (Fig. 90) may well serve as an example.

Coils *1* of steel sheet are stacked on base *2* and covered with double-walled retort *3*. Then furnace *4* is placed over the muffle. A protective gas (usually nitrogen) circulates inside the retort, protecting steel

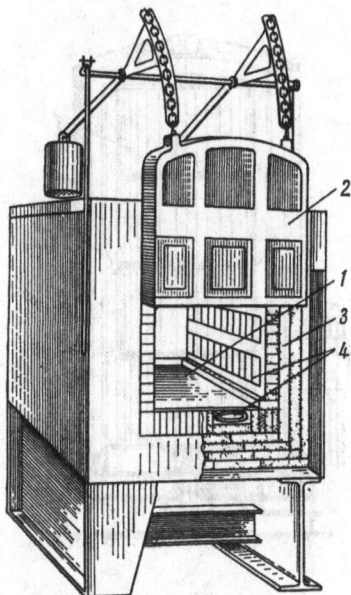

Fig. 89. Electric box furnace for heat treatment of small-size items

1 — hearth plate; *2* — door; *3* —refractory lining; *4* — heaters

against oxidation and decarburization. The furnace is heated with gas supplied to burners *5*. A rapid and uniform heating is ensured by circulating the protective gas by fan *6*. Upon annealing, the coils are cooled under the retort, the furnace bell being used to heat another batch of coils.

Steel parts are quenched in water or oil quenching tanks either located near the heating furnace or combined with the latter into a single unit; an example is a unit for processing bearing parts. These integrated units are very efficient.

Heat-treating furnaces are heated with gas, fuel oil, or electric power. Solid fuel is almost never used at heat-treating plants.

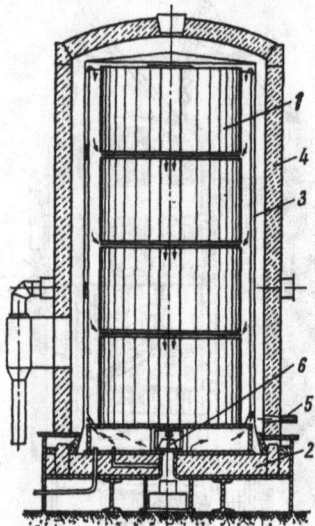

Fig. 90. Electric bell furnace for annealing coils of low-carbon steel sheet in atmosphere of nitrogen

1 — sheet coils; *2* — base; *3* — double-walled retort; *4* — furnace; *5* — burner, *6* — fan

28.6. Controlled Atmospheres

In the process of heat treatment, steel is heated to a high temperature and may remain so for long periods of time. As a result, steel surface oxidizes and carbon is burned out of it. This requires a subsequent working, such as descaling, machining, etc.

This is why heat-treatment practice seeks to eliminate or reduce the oxidizing and decarburizing action of the furnace atmosphere by creating *controlled atmospheres* of a predetermined composition.

These protective atmospheres are usually composed of carbon monoxide, carbon dioxide, nitrogen, hydrogen, and methane. The composition and application of some of the protective atmospheres are given in Table 9.

Controlled atmospheres are prepared in special installations, also sited in the heat-treatment shops. The required composition of a protective atmosphere is ensured by various means. Usually this is done by one of the following processes: decomposition of ammonia, burning of gas or charcoal, decomposition of petroleum products (fuel oil, kerosene, etc.).

The surface of parts, treated in protective atmospheres, is bright and requires no complicated finishing.

28.7. Induction Heating

Many parts of various machines operate under heavy loads and should be very tough and medium hard. At the same

Table 9

Composition of Controlled Atmospheres

Content, per cent					Application
CO_2	CO	H_2	CH_4	N_2	
—	—	75	—	25	Annealing of low-carbon and stainless steels; brazing of steels
—	3	3	—	94	Annealing of carbon and alloy steels; normalizing of alloy steels; hardening of high-carbon alloy steels; tempering of steels of all types
4-6	8-15	10-16	2	61-76	Annealing, normalizing, hardening, and tempering of carbon and alloy steels
—	18-22	36-40	2-4	34-44	Annealing, normalizing, and hardening of high-carbon and alloy steels

time, their surface must be fairly hard to resist wear. Combination of a high hardness in the surface layer with adequate toughness of the core of a steel part may be ensured through surface-hardening by high-frequency currents (10-100 kilohertzes).

A high-frequency current passing through an inductor produces a magnetic field around it. If a steel part is placed in this field, a current is induced in the part and converted into heat. A major property of high-frequency currents is that they flow preferably in the surface layer. The higher is the current frequency, the smaller is the depth of its penetration into the part.

If a heated surface layer is water quenched, it will harden. The hardness of the surface layer will increase, while the core of the steel part will remain unaffected.

The inductor is generally given the shape of the worked part to minimize dissipation of energy. For example, an in-

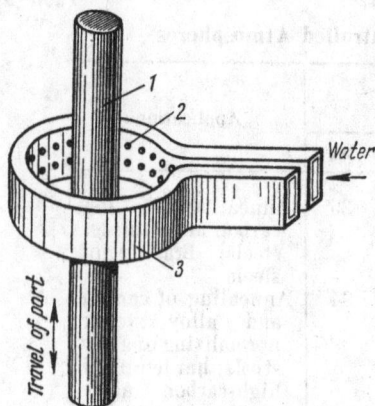

Fig. 91. Inductor for induction hardening cylindrical items

1 — heated item; *2* — holes for supplying water; *3* — inductor

ductor shaped as a ring or a spiral is used for hardening of cylindrical parts (Fig. 91). The inductor and the part may move relatively to one another when a part of considerable length (e.g., a shaft) is to be hardened. The sources of high-frequency currents are dynamo generators or vacuum-tube oscillators.

Special machines have been developed for induction hardening of mass-produced items, such as gear teeth or automobile crankshafts.

When the surface to be hardened has gained the required temperature, water is injected through apertures in the inductor or through a spraying device. Hardening may also be effected by conventional means, i.e., by immersion in a quenching bath.

The rate of induction heating is very high, exceeding 1,000 degrees (Centigrade) per minute. Therefore, the heated surface of a steel part has no time to oxidize and decarburize. Induction hardening is a highly efficient heat-treatment technique extensively used in engineering.

Induction hardening is used to process many automobile parts, such as crankshafts, spring pins, gearbox shafts, etc.

Low-frequency currents are also employed in heat treatment in applications where a steel part must be heated through.

CHAPTER 29

Heat Treatment of Steel

Steels are subjected to preliminary and final heat treatments. Preliminary heat treatment is aimed at relieving internal stresses, reducing hardness before machining, and preparing the structure of steel for a final heat treatment.

Internal stresses, excessive hardness, and incorrect structure are inevitable in the manufacture of steel items by casting or forging.

If a casting has sections of different thickness, the internal stresses will unavoidably arise at their joints due to different cooling rates. If a casting is bulky, the chemical compositions (and, therefore, the structure) of its internal and external parts may differ. When a steel contains much carbon or alloying elements, it may harden even on cooling in the air, and thus become hardly machinable.

In all these cases the defects may be corrected by appropriate heat treatments.

The final heat treatment imparts the steel part the strength, plasticity, and other properties required for the service. After the final heat treatment and some finishing to size (grinding, etc.), the steel part is ready for service.

29.1. Heat Treatment of Steel Castings

Steel castings are manufactured from casting grades of carbon and alloy steels.

Steel castings are usually coarse-grained, and, therefore, insufficiently strong. In addition, they are strained by internal stresses which may cause cracking. This is why castings are commonly annealed or normalized and tempered.

Only plain castings from low-carbon steel intended for non-critical applications are not subject to heat treatment.

Castings may be annealed and normalized (preliminary heat treatment) in any type of furnace capable of providing required temperatures. Upon holding for a time necessary for a thorough heating, the castings are cooled: in annealing, together with the furnace down to 300-400°C, then in the air; in normalizing, in the air.

Normalized castings are somewhat harder and stronger than the annealed ones.

The final heat treatment of castings is effected to obtain required strength. This is usually achieved by normalizing and tempering or hardening and tempering.

Castings are most often oil-quenched; plain castings may be water quenched. Cooling after tempering is in the air or water.

29.2. Heat Treatment of Rolled Stock and Forgings

Rolled stock from steels containing up to 0.4 per cent carbon are commonly air cooled in stacks or packs, thereby being practically subjected to normalizing. A hot-rolled steel is usually fine-grained and has good machinability.

If the size of the grain must be reduced still further, thus improving toughness and strength (for example, those of sheet), the steel is additionally normalized or hardened and high-tempered. If it is necessary to lower the hardness only to facilitate machining, a carbon steel part is tempered.

Furnaces for heat-treating rolled stock are usually sited at the iron and steel works after the rolling mills. The furnaces should have an adequate throughput capacity. On an ever increasing scale in recent years, carbon and low-alloy steels are heat-treated directly from rolling heat, which eliminates the need for costly heating furnaces and additional heating.

Structural alloy steels are subjected to high tempering or annealing. The main purpose here is to lower hardness, as alloy steels harden during post-rolling air cooling. Besides, some alloy steels may have internal cracks due to a high content of hydrogen (the so-called *flakes*). This defect may be prevented by a special cooling technique subsequent upon rolling or by annealing.

For example, rolled ball-bearing steel more than 60 mm in diameter is cooled after rolling in special pits to the following schedule: minimum charging temperature 700 °C, cooling period 24 hours, maximum discharge temperature 150 °C.

Rolled stock from carbon or alloy tool steel is also subjected to a special heat treatment, consisting of anneal-

ing followed by controlled cooling, to obtain a fine-grained
structure with good machinability.

Stainless steels are annealed or tempered after rolling
to reduce hardness.

Carbon steel forgings, depending on the properties required of them, may be subjected to the following treatments: annealing, normalizing, normalizing and tempering, hardening and tempering.

Heat treatment of alloy steels is more complicated as it
usually requires control of the heating rate; annealing and
tempering of alloy steels require additionally control of
the cooling rates. After hardening and a high tempering
with water cooling, it is sometimes necessary to conduct
an additional tempering at 450 °C to relieve internal stresses.

29.3. Heat Treatment of Machine Parts

Final heat-treatment procedures for machine parts are
most diversified. It is the final heat treatment that gives
the part the properties specified by the designer. Necessary
though the preliminary heat treatment may be, it only
prepares the structure for the final heat treatment or facilitates the machining of the part.

The type of final heat treatment depends on the kind
of part manufactured from a given steel, and on its future
operating conditions. For example, grade 45 steel is used
for the manufacture of gears, shafts, axes, but each of these
items requires a different heat treatment.

Therefore, we shall discuss the kinds of heat-treatment
procedures for some carbon and alloy steels with due regard
for their applications.

Grade 20* *steel* may be used in non-critical applications
(fasteners, bushings, levers, etc.) with no heat treatment.

When heat-treated, this steel is used for the manufacture of piston pins and spring pins, small-size gears, etc.
These parts are carburized at 910-920 °C, hardened from
860-870 °C in oil, and tempered at 170-200 °C. As a result,
the surface becomes hard (Rockwell hardness 60) and wear

* Designations of Soviet-made steels are deciphered in Chapter
5, pp. 38-41.

resistant, the core remaining soft (Brinnel hardness 160-180) as the steel is case-hardened only.

Grade 45* steel is used for the manufacture of high-strength parts (gears, spindles, crankshafts, track pins, piston pins, etc.).

When the part must be strong throughout, the heat treatment consists in water or oil quenching from 830-840°C and tempering to the required hardness.

When the surface only must be hard (Rockwell hardness 60) and wear-resistant, induction hardening and a low tempering are applied. This is the technique used for piston pins, spring pins, shafts, axles, and other automobile parts.

Grade 40X* steel is similar in application to grade 45 steel, but is stronger and tougher. This steel is heat-treated similarly, but the hardening temperature is somewhat higher (830-860°C).

Grade 18XΓT* steel is used for the manufacture of heavy-duty gears, shafts, axles, etc. The parts are first carburized at 900-920°C, then oil hardened from 780-800°C and tempered at 200°C. This results in a high strength and toughness of the core combined with the greatest possible hardness of the surface.

Grade У7* carbon tool steel is used for the manufacture of wood-working tools (drills, routing cutters, saws), chisels, hammers, dies, etc.

This steel is heat-treated by hardening from 780-830°C in oil or water, or successively water and oil (interrupted quenching), then low tempering at 160-180°C to Rockwell hardness 60-63.

Grade P18* steel is a high-speed tool steel. In addition to carbon, it contains 18 per cent tungsten, 4 per cent chromium, and up to 1.4 per cent vanadium. P18 and P9 (similar to P18) steels are widely used for the manufacture of metal-cutting tools, such as lathe tools, milling cutters, taps, drills, etc.

An advantage of rapid steel is that it retains a high hardness at high temperatures attained by the tool in operation, so that its cutting properties are unaffected.

---

* Designations of Soviet-made steels are deciphered in Chapter 5, pp. 38-41.

To obtain maximum hardness, P18 steel is treated as follows. First it is heated to 1,260-1,280°C, the heating to 800-850°C being done very carefully and slowly, then at a faster rate. This is ensured by heating the part in two different furnaces or baths. Frequently used are salt baths, as for the high-speed steels it is particularly important that no decarburization of the surface occurs.

The tool is quenched in oil or in molten salt at 500-550°C for a short time, then air cooled. It acquires a Rockwell hardness up to 63, and this is not the limit. A higher hardness is achieved by triple tempering at 560°C. In tempering, small particles of hard carbide (compounds of alloying elements with carbon) precipitate in the structure of the steel, this increasing the Rockwell hardness to 65. Greater hardness can be obtained by case-hardening which involves the saturation of the steel surface with nitrogen and carbon.

Grade X18H9T* stainless steel is used for parts operating at high temperatures or in weak corrosive media in various industrial applications, such as manufacture of nitrogen or paints and varnishes, food industries, etc. This steel has a low post-rolling corrosion resistance, and to improve it the parts should be hardened from 1,050-1,100°C in water or in the air.

Grade EX9K15M* steel is used for the manufacture of permanent magnets. To obtain maximum magnetic properties it is hardened from 1,040°C after a 10-minute soaking at the temperature. It is cooled in the air until magnetic properties appear (as evidenced by a horse-shoe magnet test), then in oil.

To stabilize the magnetic properties, the magnets are kept for 15-25 hours in boiling water.

As is seen from the above examples, heat treatment of steels depends not only on their compositions, but also on their applications.

The effect of heat treatment upon the various grades of steels is given in reference manuals to aid the designer in choosing the required grade of steel.

---

* Designations of Soviet-made steels are deciphered in Chapter 5, pp. 38-41.

CHAPTER 30

Alloy Steels

30.1. Applications of Alloy Steels

Alloy steels are used industrially no less extensively than carbon steels are.

The harder a carbon steel, the lower its ductility and toughness. A steel hardened to martensite is very brittle and cannot withstand impacts. In contrast, an alloy steel may combine high hardness and strength with high toughness and ductility. This is the chief advantage of alloy steels.

Besides, if a bulky part from carbon steel is hardened, it will not harden throughout and its properties will not be uniform across its section. An alloy steel, however, is capable of hardening throughout the same cross section even when cooled in oil, salt, etc. The internal stresses will be smaller than on hardening in water. This is another advantage of alloy steels.

Finally, as has been mentioned above, alloying elements can impart specific properties, such as oxidation resistance, acid resistance, magnetic properties, etc. Alloy steels are more expensive than carbon steels, and so they are used only when carbon steels fail to ensure the required properties.

30.2. Effect of Alloying Elements Upon Steel Properties

Alloying elements not only make a steel stronger and impart to it special properties, but also affect the transformations in its structure on heating and cooling. In particular, they alter the position of critical points A_1 and A_3; in other words, they influence the configuration of the iron-carbon diagram (Fig. 8).

This should be taken into account when choosing heat treatment temperatures (see Fig. 87) for a steel containing alloying elements. Let us discuss the influence of some of the latter.

Chromium lowers the A_3 point and raises the A_1 point; it sharply increases the hardenability and prevents grain growth in heating (which is likely to occur in carbon steels); it enhances resistance to corrosion, oxidation, and wear; delays loss of strength at high temperatures. Chromium is completely soluble in ferrite.

Nickel lowers the A_1 and A_3 points; its influence upon grain growth is small, but it enhances hardenability; it raises both strength and toughness. Maximum solubility in ferrite is 25 per cent.

Manganese lowers the A_1 and A_3 points; it accelerates grain growth in heating, increases hardenability and wear resistance, and neutralizes the adverse influence of sulfur. Maximum solubility in ferrite is 12 per cent.

Silicon raises the A_1 and A_3 points; it accelerates grain growth in heating, increases hardenability and magnetic permeability, and binds chemically the harmful oxygen contained in steel. Maximum solubility in ferrite is 18 per cent.

Aluminium raises the A_1 and A_3 points; it retards grain growth, increases the coercive force and hardness and binds oxygen and nitrogen. Its maximum solubility in ferrite is 30 per cent.

Molybdenum raises the A_1 point and lowers the A_3 point; it delays grain growth and loss of strength in heating and increases hardenability and coercive force. Its solubility in ferrite is 5 per cent.

Copper lowers the A_3 point, increases the strength, hardenability, and resistance to air corrosion. Its solubility in ferrite is 0.4 per cent.

Tungsten raises the A_1 and A_3 points, delays grain growth and loss of strength in heating, enhances hardness and hardenability. Its solubility in ferrite is 6 per cent.

Cobalt raises the A_1 and A_3 points; it reduces hardenability and increases hardness and coercive force. Its solubility in ferrite is 80 per cent.

It can thus be seen that each of the above elements affects the properties of steel in a specific manner. This is why a steel is often alloyed with several elements, particularly when it is desired to obtain a combination of different properties, for example, a good hardenability and a high corrosion resistance. There are steels alloyed with three, four, and more elements.

30.3. Classification of Alloy Steels

As each of the alloying elements affects the transformations in steel in its own specific way, the structure of different alloy steels cooled at the same rate is bound to be different, too.

Steels are most often classified according to the structure they have after air cooling.

Steels of the pearlite class after air cooling (i.e., normalized) have a structure composed of ferrite and pearlite, like that of carbon steels. In contrast to carbon steels, the ferrite of alloy steels carries atoms of alloying elements in addition to atoms of iron.

Pearlite class alloy steels are the most widely employed ones. They are always heat-treated, most frequently by hardening and tempering.

Steels of the martensite class have, upon normalizing, a martensitic structure, i.e., they harden in the air.

Steels of the austenite class, when cooled in the air, suffer no transformations and their structure remains austenitic irrespective of heating or cooling.

Steels of the ferrite class have a ferritic structure when normalized and also suffer no structural transformations.

This classification of alloy steels is conventional, since martensitic structure can be obtained in pearlite class steels by appropriate cooling methods. This is why steels are often classified according to their application into tool, structural, scale-resistant, etc. classes.

REVIEW QUESTIONS

1. Why is heat treatment necessary and what are the main kinds of heat treatment?
2. What structural transformations occur in steel in the process of heating and cooling?
3. How can heat-treatment temperatures be determined from the iron-carbon equilibrium diagram?
4. What kinds of annealing treatments are known and what do they involve?
5. What is hardening and tempering of steel? What methods of hardening are there?
6. Describe box, conveyor, and bell furnaces and state their destination.

7. What is the purpose of controlled atmospheres?
8. How are high-frequency currents used for heat treatment?
9. What heat treatment is applied to steel castings?
10. What kind of heat treatment is employed for rolled stock and forgings?
11. Speak on the final heat treatment of carbon-steel tools.
12. What are the advantages of alloy steels as compared to carbon steels?
13. What is the influence of chromium, nickel, manganese, and silicon as alloying elements?
14. How are alloy steels classified by their structure?

INDEX